D0398405

# The Nine Numbers of the Cosmos

# *The* Nine Numbers *of the* Cosmos

Michael Rowan-Robinson
*Imperial College, London*

Great Clarendon Street, Oxford OX2 6DP

Oxford University Press is a department of the University of Oxford.
It furthers the University's objective of excellence in research, scholarship,
and education by publishing worldwide in

Oxford New York

Athens Auckland Bangkok Bogotá Buenos Aires Calcutta
Cape Town Chennai Dar es Salaam Delhi Florence Hong Kong Istanbul
Karachi Kuala Lumpur Madrid Melbourne Mexico City Mumbai
Nairobi Paris São Paulo Singapore Taipei Tokyo Toronto Warsaw

with associated companies in Berlin Ibadan

Oxford is a registered trade mark of Oxford University Press
in the UK and in certain other countries

Published in the United States
by Oxford University Press, Inc., New York



A catalogue record for this book is available from the British Library

Library of Congress Cataloging in Publication Data

(Data applied for)

ISBN 0 19 850444 6 (Hbk)

Typeset by EXPO Holdings, Malaysia

Printed in Great Britain
on acid-free paper by
Bookcraft (Bath) Ltd
Midsomer Norton, Avon

# Contents

*To Mary and Nicola*

## Credits for illustrations

Figs 1.1, 4.1: Anglo Australian Observatory

Fig. 1.3: National Optical Astronomy Observatory

Figs 2.1, 3.2: Astronomical Society of the Pacific

Fig. 2.3: J. J. Condon

Figs 2.4, 5.1: Bell Labs

Figs 2.5, 2.6, 8.2, 9.7: NASA

Figs 3.4, 9.6: Palomar Observatory/California Institute of Technology

Fig. 3.5: A. Riess

Fig. 3.7: A. Stockton

Fig. 4.2: A. Penny

Fig. 6.5: T. Sumner

Fig. 7.1: S. D. M. White/Virgo simulation

Fig. 7.2: J. Silk

Fig. 9.3: Lick Observatory

Figs 9.4, 9.5, 10.1, 10.2: European Space Agency

Fig. 10.3: High Energy Physics Group, Imperial College

# Prologue

## *Que sais-je?*

In 1576 the French philosopher and essayist Montaigne ordered a medallion to be struck with the words *Que sais-je?* (What do I know?) inscribed on it. He wore this around his neck for the rest of his life to remind himself that nothing should be believed without evidence.

Montaigne's motto is a stern reminder to our present age, where many people want to believe the incredible regardless of the evidence. There is a flowering of religious cults and of pseudo-science like astrology, the occult, new age theories. Even in science, it seems that to catch the reader's eye a writer should discuss how the Big Bang originated or whether worm-holes can be used to travel through time. Now these are both very interesting questions but they can't really be included amongst things that we know. No worm holes have been found and we have no direct evidence on how the universe began.

It is natural that we ask questions like 'how did the universe begin?'; 'why are we here?'; 'does the universe have a purpose?' We have an innate hunger for understanding which goes far beyond the technical business of survival. Because these questions are very difficult to make progress on, and perhaps may never be answered, it is natural to listen to those who claim to have an easy answer. Religions and pseudo-science offer answers to some of these very difficult questions, so that the questions can be put aside. But where questions are amenable to searches for evidence, for example the age of the universe or the origin of species, the answers given by religious and pseudo-scientific texts invariably turn out to be wrong. Creationism and astrology are nonsense. The Book of Genesis offers no insight into the evolution of the universe or of life. And even on questions where there are no clear answers yet from science, I personally prefer to follow Montaigne's road and live my life facing up to these difficult questions. We appear to be here in this universe by chance and possibly to be alone. I want to live and go to my grave facing the universe as it is, trying to be honest about what we do not know about it.

So in this book I want to present what we really know about the universe and also to make very clear what we do not know. Each chapter is built around a very basic fact about the universe: that we exist, that we are not in a special place, that the universe is expanding, that the universe is of finite age, and so on to more exotic facts that we have discovered during this wonderful century of astronomy and cosmology. We have to hope future generations will remember us for achievements like these rather than for the century's darker episodes.

I have encapsulated our knowledge of the universe into nine numbers, which at the moment appear to be independent characteristics of this universe, quantitative aspects that we can hope to measure precisely in the next decades. Some of these numbers, like the Hubble constant and the age of the universe, are obvious choices. Others may be less obvious, reflecting the subtlety of modern cosmology.

As we look out into the universe, we also look back in time. When we study galaxies at the limits of today's telescopes, we look back to a time about a billion years after the Big Bang, roughly 10% of the age of the universe. The cosmic background radiation that we see at microwave wavelengths gives us a snapshot of the universe about 300 000 years after the Big Bang. Analysis of the abundance of the light elements helium, deuterium, and lithium gives us a glimpse of conditions in the universe a few minutes after the Big Bang. And with the accelerators of modern experimental particle physics we are able to probe conditions of density and temperature which existed in the universe a million millionth of a second after the Big Bang. Much of the cosmology of the observable universe has therefore become a matter of quantitative study. So, for example, we are confident that the universe is expanding and are now trying to pin down the Hubble constant, which measures the rate of the expansion.

The very early universe, the beginning of the universe, remains experimentally and observationally inaccessible, shrouded in mystery, and is therefore an area ripe for speculation. Some of these speculations have achieved widespread recognition (Alan Guth's idea of inflation, Stephen Hawking's idea of a no-time boundary, Grand Unification, superstring theory) and are the subject of intensive study by the theoretical particle physics and cosmology community. But they remain speculations nonetheless. I will touch on some of these ideas but they are not the central theme of the book. In the last chapter I will return to what the prospects are for progress in these difficult and opaque areas.

Over the next 10 to 15 years we can expect quite dramatic progress in the accuracy with which we know most of the nine numbers. At the moment all but a few are the subject of controversy but many of these controversies will be resolved, for example by the MAP and PLANCK missions, which will map the microwave background radiation with impressive precision during the next decade. Some of the features characterized by these numbers may turn out either to be far more complex than we realize at the moment or not to be independent quantities, but to be derivable from other numbers using improved theories.

Why nine numbers? Could it be 10, or 100? In selecting these quantities, I am focusing on what seem to me to be the major aspects of our knowledge (or ignorance) of the observable universe. There are other quantities that cosmologists measure and try to predict from theoretical models: for example, the luminosity function of galaxies, which measures the proportions of galaxies of different luminosities found in an average region of the universe; the luminosity density of the universe at the present epoch, which measures the total amount of light emitted by galaxies per unit volume; and the average rate of star formation in the universe today. However, these are all aspects of the history of galaxies rather than the history of the universe itself. I am only trying to characterize what we know about the large-scale universe—I am not trying to describe the contents of your fridge. Even if all the quantities discussed in this book were known with immense precision, we would still not be able to predict the details of how stars and planets form, how life arises, or what is in your fridge. Phrases like 'Theories of Everything' are very misleading. They are really just 'Theories uniting the basic forces of physics'.

There are certain fundamental constants of physics, like the speed of light and the constant of gravitation, which play an important role in understanding the universe, but are already very precisely determined. The accurate measurement of these constants was one of the achievements of nineteenth-century science and the accuracy has been improved still further during the twentieth century. Gravity is the main force controlling the evolution of the universe. The universal constancy of the speed of light in a vacuum, regardless of the motion of the source or the observer, is at the heart of Einstein's special and general theories of relativity, on which modern cosmological theories are based. Despite the fundamental importance of these numbers for cosmology I have not included them among the nine numbers of this book. The book is about the open areas of cosmology and I can take the speed of light and the gravitational constant as already known.

I do not include the basic numbers of atomic and particle physics, like the mass of the electron, the fine structure constant, or the dozens of parameters of the 'standard model' of particle physics. And I do not even include the so-called Eddington numbers, which certainly could have a cosmological significance, though at present it is completely unclear what that might be. Arthur Eddington noticed in the 1930s that if you work out the ratio of the electromagnetic force between an electron and a proton to the gravitational force between them, and also the ratio between the radius of the universe and the classical radius of the electron, then both ratios come out to be almost the same huge numbers, close to $10^{40}$, 1 followed by 40 zeros. Eddington argued that this could not be a coincidence.

Do numbers rule the universe? In focusing this survey of our knowledge of cosmology, the achievements of the twentieth century, and the prospects for the future, on certain numbers, I am bound to give the impression that numbers rule the universe. Some theoreticians do take the view that the universe is deeply mathematical in its structure and that our quest is therefore to unravel this deep mathematical structure. This Platonic view, that the universe is a manifestation of some kind of ideal, mathematical form, is very fashionable today. Some of its proponents are so astounded by this insight that they are driven to a mystical interpretation. This deep mathematical structure is God, or the mind of God, or is evidence for a creator. But why isn't this insight, that the universe is deeply mathematical, sufficient in itself? The additional mystical interpretation doesn't seem to add anything. There is, anyway, an alternative to this

Platonic view, namely that we should think of mathematics as simply an invention of the human mind, which we use as a tool to model our limited perceptions of the universe. Naturally as we try to model the extreme conditions of the early universe and draw together all the forces of physics into a single mathematical structure, that structure grows ever larger, deeper, and more complex. But it is still our invention and it still does not represent any ultimate truth about the universe. This Aristotelian view, which I share, sees the universe as something we try to characterize, measure, describe. The numbers highlighted in this book are, then, a peg on which to hang different aspects of our knowledge of the universe. Many of the numbers have a rich hinterland. All are related both to our empirical knowledge (or lack of it) and to the progress we have made towards theoretical understanding.

The idea for this book arose out of my inaugural lecture at Imperial College, which had the same title as this book. At that lecture several of my cosmologist friends tried to guess in advance what my 'nine numbers' would be. None actually came up with the same list as myself, though some of the alternative suggestions can be derived from the quantities I have chosen. So there is a definite subjective aspect, in emphasis at least, to the structure of this book. I take the view that we do not know very much about what happened before the universe was a million millionth of a second old. It could well be that 100 numbers will be needed to fully describe that early phase or that the early evolution of the universe was so inevitable that only one outcome was possible. Whether the very first instants of the universe can be understood by human beings in the foreseeable future remains to be seen.

This book can also be seen as a review of the twentieth century's achievement in cosmology. Writing in the last years of the century, I want to show you what a remarkable achievement of human effort and thought this century's progress in cosmology has been. We could not have made these advances without the knowledge that had been gained over many previous centuries. But almost nothing of what we know now about the universe was known in 1900. Because I am also writing about our ignorance, I hope I am also setting the scene for the century ahead. In a century's time someone will write a very different book about cosmology. Some of these numbers will have become physical constants like the velocity of light, whose numerical value is so well known that we do not bother to think about it any more. Others will be seen to be irrelevant in the new perspective of the time. Although the boundary of our earliest definite knowledge will have been pushed back in time, perhaps by many orders of magnitude, I predict that the very earliest instants of the universe will remain as inaccessible as ever.

Acknowledgements: I thank Fadi Al-Mufti for reading and commenting on an early draft and Stephen Warren and Andrew Liddle for their corrections, which saved me from several errors. Any remaining errors or distortions are entirely the author's responsibility.

Chapter 1

# We exist

*Not how the world is, but that it is, is what is mystical.*

Ludwig Wittgenstein, *Tractatus logico-philosophicus*

We exist.

There is an extraordinary range of profound implications of this simple fact. Life exists in the universe and has evolved in such a way, here at least, that at the last second of the hour an intelligent species has emerged. Here, now, perhaps only for a fleeting instant, consciousness exists in the universe. This book tries to show what this consciousness and this intelligence has achieved in understanding the universe we humans find ourselves in. Much of this understanding has been gained in the past century, which corresponds to only a hundred millionth of the age of the universe (if we think of the age of the universe as a year, then the past century would correspond to the last one third of a second of that year). If this is the first time that such understanding has been achieved in the universe, then cosmological knowledge is a truly novel phenomenon. For me it is a wonderful moment in which to have existed and worked in the field of cosmology.

At the very least we know the universe has to be such that we could have emerged in it. And, more mundanely, we are made of a certain kind of stuff, atoms, which pervade the universe. Ah, but there is nothing mundane about this stuff we are made of because it carries the history of the universe within it. The subject of this chapter, then, is this ordinary matter, what it is, how the elements were formed. And the first of the nine numbers of the cosmos will be the average density of ordinary matter in the universe today, a quantity which seems prosaic but which turns out to have profound significance and which probes back into the early universe.

The doubts come crowding in. Do we exist? Few thinking people have not asked themselves at some point in their lives whether this whole existence could be a phantasm, an illusion. Given the fragile and distorting nature of our perceptions, how can we convince ourselves that there is something solid out there, a world, a universe, to act as backdrop for our mental drama? The more we dissect this world, these atoms, the more we find that this apparent solidity is a chimera. The table on which I write is a shimmering lattice of atomic nuclei through which a river of electrons flows, a vacuum permeated by unimaginably strong force fields. The electrons and other atomic particles, which are supposed to form the bedrock of our existence, turn out to be mere probabilistic apparitions, able to be everywhere and nowhere, attaining a definite identity only when we, the observer, make a measurement. In quantum theory the universe seems to need us not only to achieve consciousness of itself but even to have any definite reality at all. This is called 'the Copenhagen interpretation' of quantum theory, after the school of quantum mechanics founded in Copenhagen by Niels Bohr, who advocated this view. The alternative interpretation is that the different possible states of every particle exist simultaneously in an infinite number of parallel universes—the 'many worlds' view. Every decision we made was made differently in another universe. Nothing is what it seems.

Our minds, too, have no greater solidity than the table. This powerful sense of identity, of an interior conversation, is a product of the firing of billions of neurones, of the distribution and motions of certain chemicals, with some kind of association with the region of the brain known as the thalamus. The unravelling of the mechanics of consciousness remains one of the great unsolved problems of the age. But even if we knew how the brain worked when viewed from the outside—imagine a team of scientists with electrodes rigged up to a subject's head in such a way that they can study and explain the firing of every neurone in the subject's brain—we may still not be able to make the leap from the mechanics of the brain to the experience of consciousness. We can only assert that we individually feel this experience and talk to others to find out whether they have the same feeling. And always there seems to be a deeper aspect of consciousness which can not be put into words, directly at least, and which we can only find hints of in the works of the artists and poets. So, in facing the universe we can not legitimately start from our interior sense of existence as a certainty.

That way madness lies. We have to start somewhere. We exist in this universe. We look out at the world and words form in our minds, images form. There are patterns that repeat. By asking careful questions, by careful observation, by scrupulous experiments, we find that there are models, theories, explanations that make sense of much of our experience. We should not expect the whole picture to make sense nor all questions to be answerable. Life, existence, the universe, may not after all have a meaning. The scientist offers illumination of aspects of our experience and claims that there is no part of our experience that can not be illuminated by the scientific approach. But to the question 'Why am I here?', the scientist offers no answer.

For example, we can study the wonderful life that has evolved on earth. The oldest life-forms found to date are simple 'prokaryotic' bacteria 3 billion (3000 million) years old. These are almost as old as the earth itself and indeed the whole solar system, which was formed 4.5 billion years ago. The solar system is relatively young within our Galaxy, and the oldest stars in our Galaxy are 10–15 billion years old. *Homo sapiens* is at most a few million years old so we have emerged in the last seconds of the hour that is life on earth. We seem to ourselves a pretty significant species and have acquired the capability of wiping ourselves out. Before we do that we will probably wipe out many millions of other species. But the dominant life-form remains, as it has done from the start, bacteria, measured in terms of variety of species, variety of niches occupied, and possibly, in terms of total biomass (see Stephen Jay Gould's entertaining *Life's Grandeur/Full House*).

## A team of atomies

William Shakespeare, *Romeo and Juliet*, II. iv

The aspect of our existence I want to focus on in this chapter, though, is the stuff we are made of.

Well, what are we made of? Bones, flesh, blood, skin? The four humours of the alchemists? A bag of chemicals? The idea that turns out to be really powerful is that we are made of atoms. This was first suggested by a group of Ionian philosophers of the sixth century BC, especially Leucippus and Democritus, the greatest of the so-called pre-Socratics. It is a measure of the transformation made by Socrates and his disciples Plato and Aristotle that everyone who came before them is lumped together as the pre-Socratics. Yet the idea of Leucippus and Democritus, that matter, no matter how diverse in its properties and textures, is ultimately resolvable into a small number of tiny distinct particles, atoms, was one of the greatest inventions of antiquity. Ironic, then, that Plato and Aristotle should dismiss this idea and send western thought in a wrong direction for 2000 years. Interestingly, what Aristotle disliked about the atomic theory of Leucippus and Democritus was that it seemed to require an infinite number of different types of atom, one for every different substance in the universe. In *De caelo* ('On the Heavens') Aristotle argued that his theory of four basic elements, earth, air, fire, and water, had much more predictive power. Today we believe there are 92 naturally occurring elements, which is a pretty large number compared with Aristotle's four. But Aristotle could claim to have had the right idea in demanding a finite number of basic elements, although his theory only really had descriptive rather than predictive power. And when Aristotle's ideas were rediscovered in the European Renaissance, it was the alchemists who dictated thought about the nature of the elements for several centuries. When in the eighteenth century the small part of alchemy which made sense began to be identified and became the beginnings of chemistry, the idea of the atom became paramount again. When wood burns, humours, vapours, and phlogiston have nothing to do with the case. Two atoms of the element oxygen combine with an atom of the element carbon to make a combination of atoms bound together, a molecule of carbon dioxide.

## The periodic table of elements

The revival of the atomic theory is usually attributed to John Dalton (1766–1844), though a group of earlier chemists like Joseph Priestley (1733–1804) and Antoine Lavoisier (1743–94) made crucial preparatory breakthroughs. Once the idea was there, the race was on to identify the basic atoms of matter, or elements. And to try to group them together according to their chemical and physical properties. Sodium and potassium are rather similar, very reactive metals. Fluorine and chlorine are pungent reactive gases which combine with water to make acids. Gradually the structure that makes up the periodic table of the elements, first formulated in full by the Russian Dmitri Mendeleev in 1869, began to emerge. Today we know 92 naturally occurring elements on earth. The atoms are basically listed in order of their weight per atom, with hydrogen, the lightest atom, as number 1, and uranium as number 92. An atom of uranium weighs about 238 times as much as a hydrogen atom. The ordinal number (1–92) is known as the **atomic number** and the ratio of the weight of the atom to the weight of a hydrogen atom is called the **atomic weight**. So carbon has atomic number 6 and atomic weight 12.0111 and oxygen has atomic number 8 and atomic weight 15.9994. Beyond uranium there are the transuranic elements, which have been brought into a brief existence by physicists using nuclear reactors and accelerators, and which may have similar transient existences in supernovae and other violent astrophysical phenomena. They are extremely short-lived and can exist for only a tiny fraction of a second under terrestrial conditions before undergoing radioactive decay (of which more shortly).

## The structure of the atom

Now some of these atomic weights seem to be pretty much whole number ratios, like carbon and oxygen, given above. However, if we turn to chlorine, the atomic number is 17 and the atomic weight is 35.453. In fact if we could isolate pure versions of the atoms, the ratios would be close to whole numbers. For in the first half of the twentieth century we began to see what constitutes the difference between the different types of atoms. The chemical properties of the atoms are driven by a cloud of particles surrounding each atom, so light that it contributes virtually nothing to the weight of the atom. These particles are called **electrons** and were first isolated by J. J. Thomson in 1897. The atomic number is simply the number of electrons in this cloud around the atom. Not only do electrons drive all chemistry, they are also at the heart of electricity, for an electric current is simply a flow of electrons from one atom to another in a metal. Electrons carry a small negative charge, so why don't the electrons in this cloud around an atom repel each other and escape? The answer is that they are held to the atom by the core of the atom, the nucleus, which contains the main mass of the atom and has an exactly equal positive charge that holds the electrons in orbit around it by electrostatic attraction. The particles carrying the positive charge are called **protons**—they carry exactly the same charge as an electron but have about two thousand (1836.12, to be exact) times the weight. So we could in theory have atoms consisting just of a nucleus of protons surrounded by a cloud of electrons (this was in fact Ernest Rutherford's proposal of 1911). But we know of only one example: hydrogen with a nucleus of one proton and an electron cloud with just one electron. Hydrogen is the most common element in the universe, a hundred times more abundant than everything else put together, apart from helium, which is about one third as abundant as hydrogen.

When we look at all other atoms, we find that the protons generally account for less than half the weight of the nucleus. This is because there is another nuclear particle, the **neutron**, discovered by James Chadwick in 1932, which as its name suggests is electrically neutral but has a mass about the same as the proton. Table 1.1 shows the first ten elements of the periodic table. The first row gives the name of the atom; the second line gives the atomic number, which is also the number of electrons or protons. The third row gives the number of neutrons in the most common form of the element and the next row gives the total number of nuclear particles (protons plus neutrons) or **nucleons**. Now, whereas the number of protons has to be the same as the number of electrons to keep the atom electrically neutral, the number of neutrons is a bit arbitrary and in fact there can be several versions of the same element each with different numbers of neutrons. These are called **isotopes**. There are two isotopes of hydrogen, deuterium with one extra neutron and tritium with two extra neutrons. Generally, when atoms have extra neutrons they are radioactively unstable and the extra neutrons gradually change into protons, emitting an electron in a process known as $\beta$-radioactivity or $\beta$-decay (the name comes from the days when radioactivity had been discovered but not understood). Line 5 of Table 1.1 shows some of the common isotopes of these first ten elements. Because some of these isotopes can occur naturally, atomic weights of elements on earth do not always come out to be whole numbers. Chlorine on earth consists of two different isotopes, one with 35 nucleons (17 protons and 18 neutrons) and one with 37 nucleons (17 protons and 20 neutrons), in

**Table 1.1** Periodic table for first ten elements

| Name | Hydrogen | Helium | Lithium | Beryllium | Boron | Carbon | Nitrogen | Oxygen | Fluorine | Neon |
|---|---|---|---|---|---|---|---|---|---|---|
| No. of protons in nucleus (= no. of electrons in orbit) | 1 | 2 | 3 | 4 | 5 | 6 | 7 | 8 | 9 | 10 |
| No. of neutrons in nucleus (most stable form) | 0 | 2 | 4 | 5 | 6 | 6 | 7 | 8 | 10 | 10 |
| Total no. of nucleons $A$ = symbol | 1<br>H | 4<br>He | 7<br>Li | 9<br>Be | 11<br>B | 12<br>C | 14<br>N | 16<br>O | 19<br>F | 20<br>Ne |
| Abundance relative to hydrogen near sun (by mass) | 1 | 0.34 | tiny | tiny | tiny | 0.004 | 0.0013 | 0.011 | small | 0.0017 |
| Common isotopes (more or less neutrons) | $^{2}$H, $^{3}$H | $^{3}$He | $^{6}$Li | | $^{10}$B | $^{14}$C $^{13}$C | $^{16}$N $^{15}$N | $^{18}$O $^{17}$O | | $^{22}$Ne |
| | | LIGHT ELEMENTS... | | | | HEAVY ELEMENTS... | | | | |

The only other element with an abundance greater than one thousandth of that of hydrogen is element 55, iron, with an abundance relative to hydrogen of 0.0023.

proportion roughly 77.3:22.7, resulting in a net atomic weight of $0.773 \times 35 + 0.227 \times 37 = 35.453$.

There are subtle and important reasons why even when we have a pure isotope of an element, the atomic weight of the isotope is not an exact multiple of the mass of a hydrogen atom. Firstly, the mass of the neutron is 0.08% higher than the mass of a proton plus an electron. The significance of this will emerge when we come to consider an elusive particle called the neutrino in Chapter 5. Secondly, in some atomic nuclei the nucleons are tightly bound together, in others they are more loosely bound. For example, the iron nucleus is very tightly bound, and so very stable, helium is quite tightly bound, but deuterium has a rather loosely bound nucleus. How tightly the nucleus is bound determines how easy it is to break up the nucleus when two nuclei collide in a nuclear reaction. So the degree of binding is part of the nucleus's energy budget, and since energy and mass are equivalent to each other (Einstein's famous equation, $E = mc^2$), this affects the atomic weight of the atom. Because four atoms of hydrogen weigh a bit more than an atom of helium, there is spare energy when four hydrogen atoms are fused together to make a helium atom, in the process known as thermonuclear fusion. It is this spare energy which powers the sun and hence all life on earth.

So the answer to the question, 'what are we made of?', is protons and neutrons, the heavy particles, or **baryons**, and electrons, the light particles, or **leptons**. We'll find out later that there are other types of lepton, and that baryons are probably not even fundamental particles but are themselves composite. But it was one of the great discoveries of nineteenth-century science that our bodies, the earth, and the sun and stars, are made of the same kind of baryonic matter, the same elements. The main question of this chapter is, then, 'how much baryonic matter is there, on average, in the universe?' We'll come back shortly to this question. Firstly we need to probe a bit more deeply into the meaning of the periodic table.

Line 4 of Table 1.1 gives the abundance by mass relative to hydrogen found in the sun and other nearby stars of similar age. In the sun 70% of the matter is hydrogen, 28% is helium. There are very small (but as we shall see, significant) abundances of the other light elements lithium, beryllium, and boron. The total abundance of all the elements from carbon onwards, known to astronomers as the 'heavy' elements, is about 2%, with carbon, nitrogen, oxygen, neon, and iron being the most common. The relative abundances of the elements on earth are similar, with one major exception: the earth has lost most of its hydrogen and helium.

The first nine elements of the periodic table are all quite distinct, but the tenth, neon, is an inert, noble gas like helium; number 11, sodium, has similarities to lithium (being a highly reactive metal); number 12, magnesium, has chemical properties similar to beryllium, and so on. So elements 2 to 9 are chemically similar to elements 10 to 17. Thereafter the pattern becomes more complicated, but there are clear families of elements, like the noble gases (helium (2), neon (10), argon (18), krypton (36), xenon (54)) and the halogen gases (fluorine (9), chlorine (17), bromine (35), iodine (53)). These patterns, first noticed by Mendeleev, were explained by the quantum theory of the atom developed by Niels Bohr, Wolfgang Pauli, and others. The patterns in the periodic table are determined by the structure of the electron cloud around the nucleus of the atom. Bohr realized that the electrons are arranged in a series of shells, while Pauli discovered the strict rules about how many electrons can be in each shell. The most basic chemical properties of the element are determined by the number of electrons in the outermost shell.

## The origin of the elements

So far Table 1.1 (and its extension to all 92 naturally occurring elements) is purely descriptive and classificatory. It doesn't tell us why things are like this. Where did these elements come from? Were they always there in the universe or were they made somehow? One of the great achievements of the period 1940–70, in fact one of the great achievements of the century, was that astronomers and physicists realized how the elements were made.

The first speculations on the origin of the elements were by Georges Lemaître, who in 1931 proposed that the universe began as a single 'primeval atom', which then broke up in a series of 'super-radioactive' decays to form the elements. As he wrote in a later article:

> *The atom world broke up into fragments, each fragment into still smaller pieces. Assuming, for the sake of simplicity, that this fragmentation occurred in equal pieces, we find that two hundred and sixty successive fragmentations were needed in order to reach the present pulverization of*

*matter into poor little atoms which are almost too small to be broken farther. The evolution of the world can be compared to a display of fireworks that has just ended: some few red wisps, ashes, and smoke. Standing on a cooled cinder, we see the slow fading of the suns, and we try to recall the vanished brilliance of the origin of the worlds.*

Lemaître's idea was studied further in the 1930s by Maria Meyer and Edward Teller (later to become famous, or notorious, depending on your point of view, as the father of the hydrogen bomb). A very important step towards understanding the origin of the elements was taken in 1938 and 1939 by Hans Bethe, C. L. Critchfield, and C. F. von Weisäcker who showed how fusion of hydrogen into helium powers the sun and other stars for most of their stellar lifetimes. Whereas Lemaître's theory was a rather general speculation about how the periodic table as a whole might originate, the work of Hans Bethe and his collaborators was a very concrete first step up the ladder of the periodic table. As we shall see, though, not all the helium on earth today can have been made in stars.

In the 1940s George Gamow developed a rival view to Lemaître's, that the origin of matter was in a very hot (billions of degrees Celsius) nuclear gas, which he called ylem, from which atomic nuclei would grow by aggregation of protons and neutrons as the gas cooled. Gamow developed this idea in collaboration with his two young associates Ralph Alpher and Robert Herman, and others, and the basic theory was published in a paper by Alpher, Hans Bethe, and Gamow (the $\alpha\beta\gamma$ theory) on April Fool's day 1948. Bethe's name had been added to the paper just to make the $\alpha$-$\beta$-$\gamma$ joke. However, in 1949 Enrico Fermi and Anthony Turkevich showed that there would be great difficulty in getting the process to work beyond lithium, element number 3. In the same year Alpher and Herman calculated the consequences of this model in more detail and estimated that the radiation which would have dominated the energy density of the universe during the early stages of the universe would today have a temperature of 5 degrees Kelvin, that is 5 degrees above the absolute zero of temperature (–273 degrees Celsius). This was, as we shall see, rather close to the mark. In 1953, with James Follin, they gave the first detailed modern discussion of the physics of the Hot Big Bang phase, but left reexamination of the formation of the elements by thermonuclear reactions to 'further study'.

The rival view on the origin of the elements being advocated during the 1940s was that they were made in stars, with the Dutch astronomer G. B. van Albada advocating red giant stars as the required location, while Fred Hoyle preferred supernovae. Gamow's comment was:

*What van Albada and Hoyle demand sounds like the request of an inexperienced housewife who wanted three ovens for cooking a dinner: one for the turkey, one for the potatoes, and one for the pie.*

In 1956 Fred Hoyle, who had already been trying to construct detailed models for massive stars, and Geoffrey and Margaret Burbidge, who were interested in the abundances of elements in stars, started to work at Caltech with Willy Fowler, whose group at the Kellogg Radiation Lab was measuring atomic 'cross-sections', which measure the effectiveness of each atomic species in nuclear reactions, for species important for nuclear processes in stellar interiors. Together they showed that the origin of most elements and their isotopes could be accounted for by nuclear reactions in stars. The Burbidges, Fowler, and Hoyle, or $B^2FH$ as the collaboration became known, showed how almost all the elements could be produced either during the normal evolution of massive stars or in the explosive events occurring at the end of the life of very massive stars, when they explode as supernovae. In particular the unstable (radioactive) isotopes are made in supernovae through the rapid capture of neutrons by nuclei, the 'r'-process. The stable isotopes are made by more gradual aggregation of neutrons, in the 's'-process (s for slow). So Gamow's cosmological housewife did in fact need several ovens.

The elements which the $B^2FH$ team could not account for were the light elements deuterium (which tends to be destroyed rather than made in stars), helium (stars make some of this, but not nearly enough), beryllium, and boron, and these were attributed to an unknown nuclear process X. In 1964 Hoyle reluctantly explored, with Roger Tayler, Gamow's idea that helium was made in the Big Bang. The Russians Yakov Zeldovich, Igor Novikov, and A. G. Doroshkevich also recalculated the Big Bang helium production and realized that there could be a detectable relic of the radiation-dominated phase of the universe which their Hot Big Bang model implied. They realized that this relic radiation would today be redshifted into the microwave band and also that there existed an instrument capable of detecting such microwave background radiation, the Echo antenna at Bell Telephone Laboratories in Holmdel, New Jersey. They turned to the Bell Labs technical journals to see what the performance of this antenna might be but misunderstood the units used in the journal papers and thought that the possibility of background radiation had been already ruled out. In 1964 Jim Peebles at Princeton had also recalculated the helium production and was predicting a microwave background radiation temperature of 10 degrees Kelvin. His colleague at Princeton, Robert Dicke, and two other young researchers, P. G. Roll and David Wilkinson, were designing an experiment to look for this when they heard about the discovery of background radiation by Arno Penzias and Bob Wilson at Bell Labs. We shall return to this story in Chapter 5.

Once the microwave background had been discovered it became worthwhile to recalculate the nuclear reactions expected during the hot phase of the Big Bang in much more detail and in 1966 Bob Wagoner worked as a research student with Willy Fowler and Fred Hoyle to do this calculation in enormous detail, with a network of 79 nuclear reactions. He showed that only deuterium, helium-3, helium-4, and lithium-7 would be made in any significant amounts during the Big Bang. It took over a decade for the excellent detailed agreement of these predictions with observations to be demonstrated, with the work of Dave Schramm's Chicago group being especially prominent in tying up the agreement for helium-3 and lithium-7. In 1972 the French physicists Hubert Reeves and Jean Audouze, together with Willy Fowler and Dave Schramm, showed that the remaining light elements, lithium-6, beryllium, and boron, were made by cosmic rays, particles moving close to the speed of light, ploughing through helium nuclei in the interstellar medium, a process known as spallation.

To summarize:

- One second after the Big Bang the universe consisted of pure hydrogen (with the protons and electrons moving around freely), neutrons, and particles that we will encounter in Chapter 5: **neutrinos**.
- Deuterium ($^{2}H$), helium-3, and lithium-7 were made in the Big Bang during the first 3 minutes (see Chapter 5).

- Other light elements (beryllium and boron) were made by cosmic ray spallation.
- Heavy elements were all made in stars: the stable isotopes by nuclear reactions in the cores of stars (slow neutron capture), the unstable (radioactive isotopes) by explosive processes in supernovae (rapid neutron capture).

## The cycle of gas, dust, and stars

So how did the elements in our body get from the stars where they were made? This brings us to the wonderful cycle of gas, dust, and stars that drives the evolution of our Galaxy today. Between the stars are diffuse clouds of gas, consisting mainly of hydrogen and helium. Most of the heavy elements in the cloud are in the form of small grains of dust, composed mainly of silicates and carbon, with sizes of between one thousandth and one tenth of a micron (a micron is one millionth of a metre), with an admixture of even smaller particles of aromatic compounds, known as polycyclic aromatic hydrocarbons or PAHs, which are found in exhaust fumes and tars. The sand on the beach is mainly ground-up silicates, so interstellar dust is basically sand, soot, and tar.

These clouds become more concentrated and denser, partly through dynamical processes arising from the aggregate effects of the gravitational pull of the stars in the Galaxy, and partly through forces which arise as the slightly electrically charged clouds move through the Galaxy's magnetic field. When the density of a cloud becomes high enough, the dust grains can shield the gas in the cloud from the penetrating effects of the Galaxy's ultraviolet radiation and the gas starts to form molecules, firstly molecular hydrogen (two hydrogen atoms bound together by a common electron cloud of two electrons), then carbon monoxide and a wealth of carbon-based molecules. Eventually these dense molecular clouds suffer some dynamical jolt and parts of the clouds start to collapse together, getting hotter as they do so. When the centre of a collapsing fragment gets dense and hot enough, nuclear reactions—the fusing of hydrogen to helium—begin and a star is born. The transformation of four hydrogen atoms to a helium atom generates thermonuclear energy and this is what keeps the star shining.

**Fig. 1.1** The spiral galaxy NGC 1097.

Initially stars fuse hydrogen to helium. When hydrogen is exhausted at the centre, the star will, if it is massive enough, start to fuse helium to carbon, nitrogen, and oxygen. A very massive star will continue to work its way through the periodic table till the centre is composed of iron. At this point the star blows up because fusion of iron absorbs energy rather than liberating it, so the central regions of the star face a catastrophic energy deficit. The core collapses to form a neutron star or black hole and the outer parts of the star are blown off in a dramatic explosion. How long stars live depends very sensitively on the mass of the star. A star like the sun lives for about 10 billion years before exhaustion of its nuclear fuels. A star 20 times the mass of the sun lives for only a few million years, while one of one tenth the mass of the sun can power its radiation by fusing hydrogen to helium for 10 000 billion years.

Stars return material to the interstellar medium, to the interstellar clouds we started with, in two ways. Firstly, many stars, including the

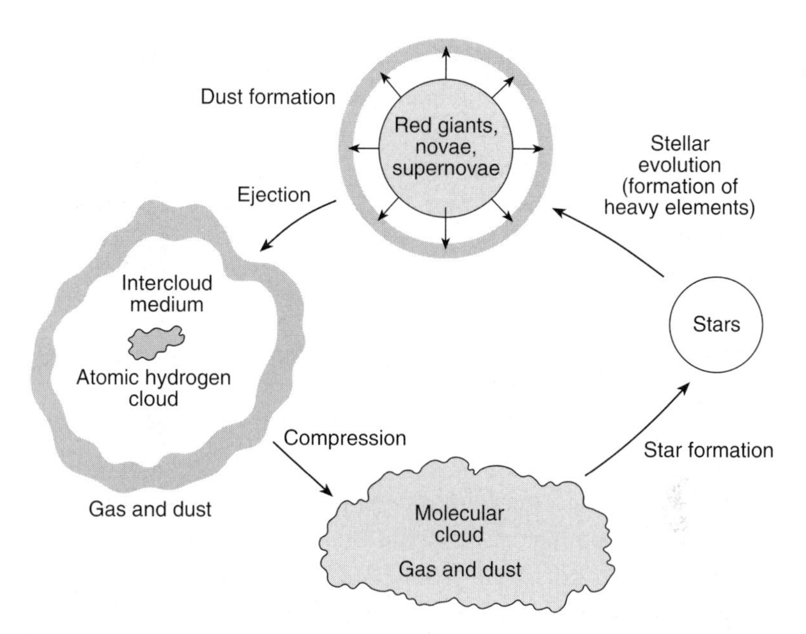


**Fig. 1.2** The cycle of gas, dust, and stars in the interstellar medium of our Galaxy. Heavy elements formed in the interior of stars are ejected in stellar winds, planetary nebulae events, or supernova explosions and condense into small grains of interstellar dust. Clouds of atomic hydrogen and dust aggregate together and, when the density is high enough, form molecular hydrogen. New stars form when the dense molecular clouds undergo compression.

sun, continuously eject gas from their surface in a steady wind. At some phases in a star's evolution this wind can be a highly prolific process. Secondly, when stars exhaust their nuclear fuels and die they often throw off a major part of the mass of the star in a violent event. So while the sun will lose little of its mass during its long hydrogen-burning phase, it will throw off a large fraction of its outer parts, perhaps half its total mass, during its dying 'red giant' phase. This happens in the form of a spectacular ejection, leaving a hot white remnant surrounded by an expanding shell of gas, a phenomenon known as a 'planetary nebula' (nothing to do with planets, though). The hot core will become a 'white dwarf' star, a dead remnant. Stars much more massive than the sun die, as we have seen, in an even more spectacular fashion, in supernova explosions in which the core collapses to form a very compact 'neutron' star or black hole and the rest of the star is ejected in a cataclysmic and dramatic explosion.

All these processes help to cycle back to the interstellar gas the elements which have been made by the nuclear reactions in stars. As the Galaxy ages, the proportion of the material in the clouds in the form of heavy elements gradually increases. Today, in the neighbourhood of the sun, this proportion is about 2%. When the solar system formed from a molecular cloud 4.5 billion years ago, the figure would have been slightly lower. The elements of our bodies were floating there between the stars, having been made in the cores of earlier generations of stars. Perhaps in 5 billion years' time when the sun becomes a red giant star, engulfing the earth, and then throws off its outer layers in a planetary nebula ejection, they will float there again.

The sun and the planets formed together and initially all shared a common distribution of elements. Jupiter and the other massive gaseous planets still share the sun's composition. However, the gravity of the lower mass inner planets (and other small rocky bodies of the solar system) was not strong enough for them to hold on to their hydrogen and helium and these have now mostly escaped.

## The origin of life—the most interesting form of baryonic matter

The earth, then, formed about 4 540 000 000, or 4.54 billion, years ago. The most interesting form of baryonic matter, life, seems to have been already in place in the form of bacteria a billion years later, judging from the microfossils in the oldest rocks found at the surface of the earth. The first mammals did not appear till 4.4 billion years after the formation of the earth and the first hominids not till 4.536 billion years after earth formed (4 million years ago). The million years of *Homo sapiens* are a minute fraction of the age of the earth and the time that human societies capable of spawning astronomers have existed, a few thousand years, is even more insignificant. What might we be capable of if we could survive for a billion years?

This brings me to the difficulty I have with the idea that advanced technological civilizations like ourselves are common in our Galaxy. Stars like the sun and planets like the earth could have formed at least 3 or 4 billion years before our sun. The abundances of the key heavy elements like carbon, oxygen, nitrogen, and iron in such systems would be very little different from those in the solar system. If people very like ourselves have been around for such enormous lengths of time, surely we would know about them? I don't find very convincing the suggestion that they keep themselves hidden so as not to interfere with our development. Some of those hypothetical predecessors would have been facing the death of their sun, as we will in a few billion years' time. Assuming that their technology has advanced a few billion years beyond ours, they would find a way to send out colonizing expeditions, as we will in due course. And they would have found us. I don't accept that there are any really fundamental limits on colonizing the Galaxy, given plenty of time. Eventually there would have been a civilization with the necessary technology and determination and they would be here. Now I know there are some crazy people around who believe that flying saucers appear all the time and that 'They' have already been here. But I think most people will agree that the history of humankind can be understood pretty well without hypothesizing any previous invasions from space. In *War of the worlds*, H. G. Wells in his usual prescient way points out a potential biological limitation to colonization, the possibility that our bacteria could be more potent than their bacteria, or at least that they may not have any resistance to our bacteria (or viruses). A truly advanced civilization facing the death of their sun and confronted by repeated failure to colonize other planets would surely put some effort into demonstrating that they had existed. There would be the astronomical equivalent of the pyramids, some kind of beacon signalling forever. While our searches for signals in the aether have been limited by our still relatively primitive technology, it seems surprising that the Galaxy is so quiet.

So there is a clear possibility that planets like the earth with bacterial life are common in the Galaxy and the universe, but that we are the first intelligent technological species to have emerged. Many biologists, from Jacques Monod to Stephen Jay Gould, are sceptical that the evolutionary path that led to mammals and to humans could ever be repeated (see, for example, Gould's *Wonderful life*; however, for a more equivocal view about the likelihood of the emergence of intelligent species in our Galaxy see Richard Dawkins' *The blind watchmaker*).

Life may not even have originated on earth. Meteorites generated by violent impacts transport material from planet to planet, especially in the direction inwards towards the sun. Bacterial life could have formed first on Mars and then travelled here. Quite complex molecules are known to form in interstellar clouds, including the tar-like polycyclic aromatic hydrocarbons (PAHs). The Orgueil meteorite, which fell to earth at Orgueil, France, in 1864, was found in 1961 to contain many complex organic compounds including all the known amino acids, probably formed in a dense cloud of molecular gas during the early stages of the formation of the solar system. Whether interstellar processes are capable of making the step to proteins and DNA remains to be seen. Fred Hoyle and Chandra Wickramasinghe pushed this kind of idea to the limit in *Life cloud*, but ran into ridicule with their suggestion that the main ingredient of interstellar dust is the bacterium *E. coli* (astronomers are reasonably certain that the dust consists of particles of silicates and carbon).

## How much baryonic matter is there in the universe?—the first of the nine numbers

The baryonic matter in the universe exists in the form of human beings and other life, planets, comets, stars, interstellar gas, and dust clouds—in a word, galaxies. In addition to the stars and gas clouds which we can see, our Galaxy almost certainly contains a large proportion of matter which has proved much harder to see, dark matter. We will return to this question in later chapters, but one form of dark matter which is relevant to our census of baryonic matter is brown dwarfs, gaseous objects with mass between that of Jupiter and the lowest possible stellar mass, about 80 times the mass of Jupiter or 0.08 times the mass of the sun. Objects in this mass range (1–80 Jupiter masses) can never get hot enough in their cores to fuse hydrogen and become stars, and are known as brown dwarfs. They are not really brown, and in fact radiate at infrared wavelengths. They have proved rather hard to detect but several examples have been found in recent years by searching for faint, very red objects in nearby star clusters. They probably do not make up a major proportion of the universe's baryonic matter.

Galaxies are often grouped together into clusters, with memberships ranging from tens to thousands. In the richest clusters we often see a huge cloud of very hot gas permeating the whole cluster. The gas is so hot, 100 million degrees Celsius, that it emits strongly at X-ray wavelengths and this is how it was found, with X-ray telescopes on spacecraft (starting with the Uhuru mission in 1970). There seems to

**Fig. 1.3** The Centaurus cluster of galaxies.

be about as much matter in the form of this gas as there is in the galaxies of the cluster. Clouds of gas have also been found in intergalactic space, through their absorption of ultraviolet radiation. It is quite hard to estimate what the total contribution of intergalactic gas clouds is to the baryonic mass budget of the universe. It could be well over 90%.

If we could only estimate the average density of baryonic matter in the universe by adding up all the above contributions, then we would end up with a very uncertain number. But it turns out that the density of baryonic matter is one of the few cosmological quantities that we can determine reasonably accurately. This comes purely from measuring the abundances of the light elements, deuterium, helium, and lithium, which were made during the Big Bang. The formation of these light elements depends very sensitively on the average density of baryons in the universe and so by measuring the primordial abundances of these elements, that is the abundance before stellar and other processes started to modify these abundances, we can determine the density of baryons in the universe. We will see in Chapter 5 that to account for the abundances of the light elements we need the average density of baryons in the universe today to be $2.5 \times 10^{-28}$ kg $m^{-3}$, to an accuracy of about 20% either way. It is difficult to digest the fact that the universe has such a low average density. For comparison, the average densities of the earth, the sun, and the human body are all within a factor of a few of 1000 kg $m^{-3}$. The density of air is about a thousand times lower than this. But the universe, 4 thousand million million million million times less dense than air, is on average a far better vacuum than the most perfect vacuum that can be manufactured on earth. The universe is an extraordinarily empty, dark, cold place. The average density in our Galaxy is about a million times higher, but still very empty relative to terrestrial densities. These low densities illustrate the enormous distances there are between stars within a galaxy and between galaxies.

This density seems very low, but it turns out to be ten times larger than the average density that exists in the form of stars in galaxies. So where are the rest of these baryons? Gas in galaxies accounts for only about 10–20% as much mass as the stars, so is of little help. Some could be due to underluminous objects like brown dwarfs in the halos of galaxies. Some could be in the form of dead stellar remnants like white dwarfs, neutron stars, and black holes. For example, we shall see in Chapter 6 that there could be a population of white dwarf stars in the halo of our Galaxy, contributing as much as 20% of the mass of the halo. This population could account for twice as much baryonic matter as is seen in the disc of stars and gas. However, most of the unaccounted for baryonic matter has to be in gas clouds between the galaxies. Individual galaxies may be surrounded by extended clouds of hydrogen (and helium) gas at temperatures of 100 000 degrees. In rich clusters of galaxies, as we saw above, we know from X-ray evidence that there is generally a huge cloud of gas at a temperature of 100 million degrees pervading the cluster, containing as much mass as is seen in clusters. And between the galaxies and clusters there is evidence for a myriad of cold gas clouds, material that never did manage to form into galaxies. The proportions in these different forms remain rather uncertain.

## Aside on units

We have been talking about kilograms, metres, seconds, years. These all seem to be highly geocentric units of measurement. The metre is essentially a human pace. The gram is the weight of a cubic centimetre of water. The year is the length of time for the earth to go round the sun and the second is defined by dividing the day into 24 hours, and the hour into 60 minutes of 60 seconds (60, because this was the base of the Babylonians' measurement and number system). Surely it is unsatisfactory to be characterizing the universe in these earth-based units? How are we going to communicate this knowledge to extraterrestrials when we meet them?

What we will try to do as the book progresses is to give these numbers in a dimensionless form, free of units. However, we are not quite ready to do that yet, so for the moment we leave the first of the nine numbers in earth units.

In fact it turns out that this density, which seems so low in terrestrial units, is just about right for us to have evolved in the universe. We would not have existed if things had been very different. If the density had been very much higher, the universe would have finished its expansion phase and have already collapsed together into a final Big Crunch without ever making galaxies, stars, planets, life. And if the density was very much lower, galaxies would still not have formed and we would again not be here. And we are only just here after all. If we think of the age of the universe as an hour, *Homo sapiens* has existed for only about one second, and has seriously studied cosmology for only a ten thousandth of a second. A million years ago, a mere instant in the universe's history, there was perhaps no intelligent species in the universe. Amazing that we exist and live at a time when we have begun to understand the universe.

Chapter 2

# We are not in a special place

*Three rocks, a few burnt pines, a desert chapel,*
*And higher up*
*The same landscape, repeated, begins again.*

George Seferis, *Mythistorema*

When we look at the other barren planets of the solar system, earth, with its teeming life, is certainly a special place. The silence of the universe may tell us that earth is unique, that no other intelligent species exist. So we may well be in a special place, the place where we are.

But from an astronomical perspective we are not in a special place. The sun is not a special star and falls in the middle of the range of stellar masses (one tenth to one hundred times the sun's mass). We are neither at the centre nor the edge of our Galaxy and our Galaxy is a typical spiral galaxy, though a fairly large one. The 'Local Group' of galaxies in which our Galaxy lies is a very typical small galaxy group, containing about 20 galaxies within a region 3 million light years across. We find ourselves neither in the core of a rich cluster of galaxies, with thousands of galaxies packed into the same volume as is occupied by the Local Group, nor in one of the vast voids between the clusters, often hundreds of millions of light years across. And as we look outwards to the vast distances reached by modern telescopes, we see the same landscape endlessly repeated: small groups of galaxies, clusters, voids. We do not lie at the centre of the universe and in fact the universe seems extraordinarily smooth and featureless once we start to look on a large enough scale, on the scale of a thousand million light years or more, the scale on which galaxies become like a grain of sand in a desert. This smoothness, uniformity, homogeneity of the universe on the large scale is one of the great surprises of twentieth-century cosmology. The second of our nine numbers will be the accuracy to which, on the large scale, the universe is homogeneous.

## From the geocentric view of Aristotle to the Copernican universe

I do not find it at all surprising that the geocentric view of the universe prevailed in ancient cultures, that the earth-centred view of the cosmos dominated the thinking of peoples as sophisticated as the Greeks and the Chinese. In some ways what is surprising is that in both those great civilizations the view arose and was debated that the earth is not in a special place and that it is just one world among many. Aristarchus of Samos, in the third century BC, proposed that not only does the earth rotate on its axis each day but also it orbits the sun, but this view did not prevail. Aristotle had considered the question quite carefully and found it easier to believe that the stars whizzed round the sky every day than that the solid earth itself, the centre of everything, could be revolving. So the earth-centred view, developed in detail by Ptolemy in the second century AD, dominated European thought until the time of Copernicus and was taken extremely literally in all attempts to explain the motions of the sun, moon, planets, and stars around the sky.

Copernicus, aided probably by Arabic and other predecessors who began to discuss again the ideas of those ancient Greek dissenters, cut through the fog of the Ptolemaic system to place the sun at the centre of the solar system. Within 50 years of the publication of Copernicus's *De revolutionibis* in 1543, Giordano Bruno in Italy and Thomas Digges in England had taken the Copernican revolution to its logical limit and argued that the sun was a typical star in a boundless universe of stars. Although the Copernican system was accepted rapidly, especially once Galileo had demonstrated in 1608 that Jupiter and its moons formed a miniature version of the Copernican model for the solar system, the direct observational proof of the Copernican system took nearly three centuries. In 1728 James Bradley showed that the motion of the earth in its orbit round the sun resulted in a systematic seasonal change in the direction of all stars on the sky, the phenomenon of **aberration**. And in 1838 an even more direct consequence of the earth's orbit round the sun, **parallax**, the change in direction of nearby stars as the earth finds itself first on one side of the sun and then, six months later, on the opposite side, was first observed reliably by Friedrich Wilhelm Bessel in the obscure star 61 Cygni. Both these effects are rather small. The magnitude of aberration is determined by the ratio of the speed of the earth's motion to the speed of light, and stars execute an ellipse on the sky of diameter 20.47 arc seconds,* which does not depend on their distance. Parallax is a much smaller effect and depends on the distance of the star. The parallax of 61 Cygni is only 0.29 arc seconds and there are only ten star systems with parallaxes larger than this, of which Alpha Centauri, the nearest star system, has the largest parallax, 0.75 arc seconds. Measurements of this accuracy were beyond the scope of Copernicus's immediate successors like Tycho Brahe, but most astronomers of the time preferred to believe that their failure to observe parallax was simply a consequence of the stars being too distant. Tycho was unusual in arguing that the absence of parallax meant that the earth stood still and the sun went round the earth, with the other planets orbiting the sun. But the door opened by Copernicus could not be closed. We now call the idea that we are not in a special place, the **Copernican principle**.

## The island universe

In the seventeenth century, Isaac Newton began to worry, under questioning from the young cleric Richard Bentley, about the stability of an infinite universe of stars under gravity. Why does everything not fall together in one place? Newton argued that if the universe were, on average, smooth and infinite, then no one place could be singled out for everything to fall to. So we have to give Newton the credit for the idea of a homogeneous universe. His contemporary Christopher Wren, who started his career as an astronomer, speculated that fuzzy objects like the Andromeda nebula might be distant star systems like

* Angles are measured in degrees, and there are 360 degrees in a circle. A degree is divided into 60 arc minutes and an arc minute is divided into 60 arc seconds. One arc second is about the angle subtended by a one-penny or one-cent coin at a distance of a mile.

the Milky Way, so that we should think of a universe of galaxies, with each galaxy like an island in a great ocean, rather than of a universe of stars. Before the discovery of the telescope rather few 'nebulae' were known. In the north, apart from the Andromeda nebula, there are just a few star clusters which look extended to the naked eye. In the southern hemisphere there are the much more impressive Magellanic Clouds, news of which was brought to Europe by Pigafetta, chronicler of Magellan's circumnavigating expedition of 1492–93. The telescope, in the hands of eighteenth-century French astronomers like Nicholas de Lacaille and Charles Messier, and above all in the survey programmes of William Herschel, was to dramatically increase the number of nebulae known and pave the way for the universe of galaxies, Hubble's 'realm of the nebulae', that we know today. Wren's speculation was the first appearance of the island universe view, an idea which was debated for over 200 years and seemed to be all but abandoned by the beginning of the twentieth century.

In the middle of the eighteenth century, Thomas Wright and Immanuel Kant focused instead on the Milky Way and the disc-like distribution of stars which seemed to be implied by the concentration of starlight towards a great circle on the sky. In 1817 William Herschel published the results of 30 years of systematically counting the stars in different directions on the sky, which firmly established that the structure of the Milky Way is a flattened disc. Herschel spent much of his long astronomical career studying the nebulae with a series of great telescopes which he constructed himself. Charles Messier had published in 1781 a list of 103 nebulous objects which comet searchers like himself needed to know about in case they mistook them for new comets. Herschel's studies were far more systematic, and in 1786 he published a list of over 1000 nebulae. Although some nebulae, like the Orion nebula, could not be resolved into stars by even the largest telescope, many were clearly clusters of stars. Herschel at first inclined to the view that most were in fact distant systems like the Milky Way, island universes, a view that Kant had supported. However, the invention of astronomical spectroscopy by William Huggins in the 1860s demonstrated that some nebulae are clouds of hot gas and this persuaded most astronomers in the second half of the nineteenth century that the nebulae are merely part of our own Milky Way system. Interest focused instead in whether the sun lay at the centre of this Milky Way disc. Studies of the stellar distribution seemed to show that the sun did indeed lie near the centre of the Milky Way system. It was not until 1918, when the American astronomer Harlow Shapley presented his studies of the distribution of globular clusters, concentrated clusters of stars with a spherical shape which are distributed in a halo around the Milky Way galaxy, that the truth emerged. The centre of the Milky Way system lies towards the constellation of Sagittarius and the sun is in fact about one third of the way between the centre and the edge. Astronomers did not all immediately accept this and the influential Dutch astronomer Jacobus Kapteyn was still arguing during the 1920s that the sun is close to the centre of the Milky Way.

## Einstein's homogeneous universe

While astronomers were debating the structure of the Milky Way, Albert Einstein was changing the whole structure of the universe. In 1916 he published his general theory of relativity, which completely changed our understanding of the nature of gravity. Gravity is not really a force but is a consequence of the curvature of space-time induced by masses. Space is curved around a massive body so a small test particle moves in a curved orbit around the body, giving the illusion of a force acting on the particle. The theory gives much the same predictions as Newtonian gravitation in weak gravitational fields like those in the solar system, but there are a few small differences.

**Fig. 2.1** Albert Einstein.

The first triumph of Einstein's theory was explaining an anomaly which had been known for over a century in the orbit of Mercury, and which Urbain Leverrier, the co-discoverer of Neptune, had tried to explain as due to a small planet nearer to the sun than Mercury. Astronomers even gave this hypothetical planet a name, Vulcan, but searches for it repeatedly proved unsuccessful. In his popular astronomy book of 1886, *The story of the heavens*, Robert Ball devotes a chapter to 'The planet of romance'. But in a footnote he remarks that Simon Newcomb, the distinguished celestial dynamicist, had already shown that the deviations between Newtonian theory and observation could not be accounted for by a planet between the sun and Mercury. The nature of the anomaly is that the axis of the ellipse in which a planet orbits the sun undergoes a slow rotation, or precession. The effect is strongest for Mercury and amounts to a precession of 43 seconds of arc per century for this planet. To put this in perspective, the purely Newtonian effect of Jupiter and the other planets causes a precession over 20 times larger than this.

The second measurable effect is more directly related to the basic ideas of general relativity. The path of a light ray passing near the sun is bent slightly round the sun, so if stars are observed near the sun during an eclipse, their position appears to be displaced about 2 arc seconds radially away from the sun and this can be measured by comparing with their positions relative to other stars six months later when the sun is out of the way in the opposite direction on the sky. This effect was observed in Arthur Eddington's famous eclipse expedition of 1919, which instantly made Einstein famous. Today we can measure this effect more precisely by observing the direction of distant radio sources as the sun passes near their line of sight. A third effect of general relativity is that clocks slow down in a gravitational field and this has been measured by transmitting radar signals to Venus or other solar system objects and measuring the extra time that elapses before the reflected signal is detected at earth, the 'radar time delay'.

In 1917, before either of these tests had confirmed his theory, Einstein applied his new ideas to the structure of the universe. In order to make the cosmological problem tractable he made the dramatic proposal that on the large scale the universe is homogeneous, the same at every point, and isotropic, that is, it looks the same in every direction. Not only is the sun not in a special place but every place in

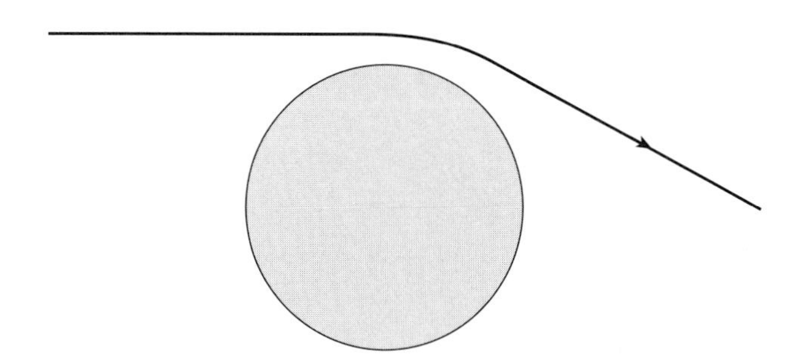

**Fig. 2.2** The bending of light. A key prediction of Einstein's general theory of relativity is that the path of a ray of light is curved by the presence of matter. The effect is shown highly exaggerated: light grazing the surface of the sun is deflected by only 1.75 arc seconds.

the universe is equivalent to every other, and every direction you look in you see the same vista. This is such a fundamental and powerful postulate that it has become known as the **cosmological principle**. Einstein was trying to construct a static, unchanging model of the universe and the assumption of homogeneity and isotropy was an essential prerequisite for this.

There could hardly have been less empirical support for this idea at the time. For most astronomers, the universe was the Milky Way system, which was clearly neither homogeneous nor isotropic. For those who thought the spiral nebulae might be distant systems, there was the problem that they seemed to be concentrated towards the poles of the Milky Way, so apparently were far from isotropic. It's not clear whether Einstein thought his model really did apply to the universe, or whether he chose these assumptions because without them he could not have found a cosmological model within general relativity. At first sight an idea like homogeneity seems impossible to check, because it would seem to involve travelling to many other locations in the universe and verifying that the universe looked the same from there.

Einstein's model was a static model of the universe and the tendency for gravity to cause everything to collapse together was counterbalanced by a new force, the cosmological repulsion, which would increase in strength with distance and so have an effect only on very large scales. As we shall see later, Einstein soon regretted adding this additional force, but the idea has made a comeback in the past

**Fig. 2.3** The isotropy of the distribution of radio sources on the sky in the northern hemisphere. (From P. C. Gregory and J. J. Condon (1991), *Astrophysical Journal Supp.*, **75**, p. 1011.)

decade. The story of the cosmological repulsion will be the subject of Chapter 8.

Soon after Einstein's announcement, the Dutch theoretical astronomer Willem de Sitter realized that there were also some non-static models for the universe—the universe could be expanding or contracting. We'll return to the issue of the expansion of the universe in the next chapter. For the moment we focus on the question of the smoothness, the homogeneity of the universe. In 1936 the American mathematician Howard Robertson, and independently the British mathematician A. G. Walker, proved a very interesting mathematical result: if the universe looks isotropic, then either we are in a special place or the universe must also be homogeneous. In an inhomogeneous universe the universe can only look isotropic from an infinitesimal fraction of the points in the universe. For example, let's take a very simple kind of inhomogeneity, where the density falls away in every direction with radius from a particular point. Now, if we are at that point then the universe will look isotropic, the same in every direction. But from anywhere else, the density will increase towards the central point and decrease in most other directions, so will not look isotropic. The great power of Robertson's theorem is that if the universe looks isotropic from earth, then it is almost certainly homogeneous and we do not need to travel to other parts of the universe to verify this.

In 1930 the American astronomer Robert Trumpler had made a very important discovery, which solved many of the problems and controversies in cosmology at the time. He realized that interstellar space, the space between the stars of the Milky Way, is filled with absorbing material, interstellar dust, which dims the light of distant objects. This explained why the spiral nebulae seemed to be concentrated towards the poles of the Milky Way galaxy. There are just as many galaxies in directions close to the plane of the Milky Way but the dimming of their light makes them much harder to see. Interstellar dust also explained discrepancies between different estimates of the distances of the nearest spiral nebulae and of the size of the Milky Way system. With this crucial insight, Edwin Hubble set out to test whether the universe of galaxies which was beginning to open up in the 1920s and 1930s could in fact be isotropic on average. He selected small fields spread all round the sky and counted the number of galaxies brighter than a given limit, making allowance for the effect of dimming by dust. However, even after this correction there were still quite large variations in the numbers of galaxies in different directions. The galaxies are not in fact distributed smoothly today but are clumped together into clusters. This result has been confirmed in a succession of galaxy surveys spanning 60 years and involving ever larger numbers of galaxies and ever deeper searches.

So it seems that the universe of galaxies is decidedly inhomogeneous, at least on the scale of these surveys. To see an underlying homogeneous universe either we must look on still larger scales today or we must probe back in time to the era before galaxies formed. Two types of galaxy survey which probe to exceptional distances have begun to show some evidence of this tendency to homogeneity on large scales. Firstly radio surveys, which tend to pick up rare and distant active galaxies that are exceptionally powerful radio emitters, usually associated with a black hole in the nucleus of the galaxy, do tend to show a rather smooth distribution on the sky. Such surveys suggest that the universe is homogeneous to at least the 10% level (fluctuations in density on large scales no greater than ±10%). And far infrared surveys based on the Infrared Astronomical Satellite's (IRAS's) survey of the sky at a wavelength of 60 microns have also reached volumes large enough (much greater than scales of 1000 million light years) to see homogeneity to the same level of precision.

## The discovery of the ripples

However, to see the true homogeneity and isotropy of the universe, and the extent to which we are not in a special place, we have to turn to a completely different phenomenon and to travel back in time much closer to the origin of the universe. In 1965 Arno Penzias and Bob Wilson, two young radio-astronomers working at the Bell Telephone Laboratories, made a discovery which was to transform cosmology. Using a huge communications antenna at Holmdel, New Jersey, to study the Milky Way at microwave frequencies, they discovered a general background radiation which was the same no matter which direction they looked in. Although they did not themselves realize at first what they had discovered, it was something that several cosmologists had been expecting (see p. 12). The microwave background radiation was in fact the relic of the Hot Big Bang era, when the dominant form of energy in the universe was radiation. The radiation they were detecting had travelled for almost the whole age of the universe before it hit their antenna.

Astronomers embarked on a campaign of study of the microwave background radiation, using ground-based radio telescopes and microwave and sub-millimetre telescopes carried on aircraft, balloons, and rockets. As the accuracy of the measurements improved it became clear that the radiation was extremely isotropic. The first small deviation from isotropy was detected in the 1970s by teams at Princeton, Berkeley, and Florence, who found that the radiation appeared slightly brighter on one side of the sky and slightly dimmer on the opposite side, by 1 part in 1000. This 'dipole anisotropy' is caused by the motion of the solar system, and of our whole Milky Way galaxy, through the radiation. Apart from this effect of our local motion it was clear by the early 1980s that the microwave background radiation is isotropic to better than 1 part in 10 000. This was evidence for isotropy 1000 times

**Fig. 2.4** Arno Penzias and Robert Wilson. (Photo courtesy of Bell Labs.)

more precise than that derived from the deepest galaxy surveys. Using Robertson's theorem we are able to say that the universe really is, to a good approximation, homogeneous and isotropic, as Einstein proposed (although strictly speaking Robertson's theorem does not tell us that almost perfect isotropy implies almost perfect homogeneity).

The spectrum of the background radiation, the spread of the radiation between different wavelengths or frequencies, appeared to have a particular simple form, reasonably close to what is called a 'Planck black body' spectrum. This form, derived by Max Planck in 1900 in the first prediction of the new 'quantum' theory, is what we expect when we look at a system where matter and radiation are in thermal equilibrium with each other. For example, if we look through a small hole into a furnace at a stable temperature, the walls of the furnace and the radiation inside it are in thermal equilibrium, with lots of interactions between the photons of radiation and the atoms of the furnace walls. The typical atom in the walls and the typical photon then both have about the same energy, which depends only on the temperature of the enclosure. In this situation the spectrum of the radiation has the characteristic form predicted by Planck, with a peak intensity at a wavelength which depends on temperature, and with the intensity falling away steeply at longer and shorter wavelengths. A body radiating with this spectrum is called a black body because it behaves like a perfectly efficient absorber and emitter of radiation. If the furnace door is opened the radiation can escape to the outside world, so the thermal balance between the matter (the furnace walls) and the radiation will be lost and the spectrum would change. Although for the first couple of decades following the discovery of the microwave background not many wavelengths could be observed, the spectrum did seem to be consistent with a Planck black body. This implied that we are looking back to an era when matter and radiation were locked together in thermal equilibrium. It became clear, as we will explore in Chapter 5, that we are in fact looking back to a phase only a few hundred thousand years after the Big Bang. At that time the matter in the universe must have been extremely smoothly distributed. How did we get from this extraordinary smoothness to the highly clumped distribution of galaxies that we see today? This will be the theme of subsequent chapters.

The story of the final proof that the microwave background does have structure, or ripples in density, when you look closely enough has been told several times. In 1989 NASA launched the Cosmic Background Explorer (COBE) satellite to map the microwave background radiation to unprecedented accuracy. After years of careful analysis the COBE team announced on 23 April 1992 that they had discovered ripples in the background and the world's media reacted with amazing fervour. Two of the COBE protagonists, George Smoot (*Wrinkles in time*) and John Mather (*The very first light*), have written very different accounts of this discovery. For how the news was received in the UK and some of the wider implications of the discovery see also my *Ripples in the cosmos.* Although there were, as often happens with big news stories, some non-ideal aspects of how the story was broken, for example George Smoot's briefing of a freelance journalist prior to the official NASA press release, and

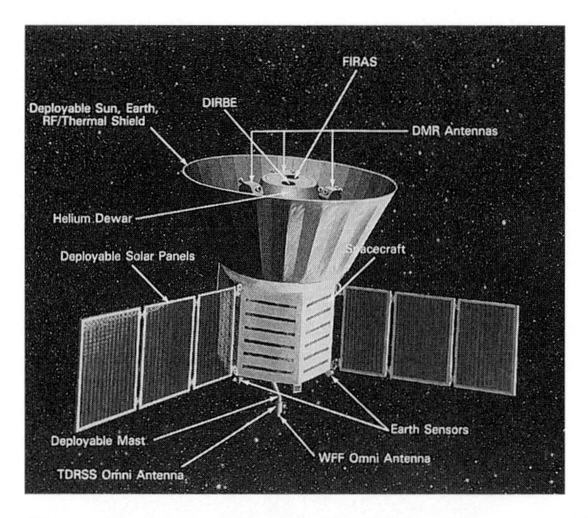


**Fig. 2.5** The Cosmic Background Explorer (COBE) satellite.

NASA's release of the COBE map of the sky with a rather misleading caption implying that the blobs on the map were the cosmic ripples (in fact most were just noise), this was an extremely important scientific discovery. George Smoot, as leader of the instrument team that made the discovery, and John Mather, as overall leader of the COBE team, both deserve full credit for this.

## The second of the nine numbers of the cosmos, the lumpiness of the large-scale universe

And so we come to the second of our nine numbers, which characterizes the accuracy to which we know the universe is smooth. The COBE team measured the average fluctuations in temperature or intensity of the microwave background radiation, denoted by $\Delta T/T$, which they found to be 1 part in 100 000, or $10^{-5}$. The telescope on COBE was quite small, so the angular resolution of the COBE observations was rather poor, 7°. If we translate this into the dimensions of the regions of universe over which we are seeing these fluctuations, back at the era a few hundred thousand years after the Big Bang, then 7° corresponds to a region 1500 million light years across. This is a very large region, about the depth of the largest of the galaxy surveys conducted to date. So it is a scale on which we think the galaxy distribution today is reasonably smooth but we do not really have much information yet. From COBE we can say that the universe was

**Fig. 2.6** The microwave background sky as seen by COBE. *Top panel*: the dipole anisotropy due to the earth's motion through the cosmic frame. *Middle panel*: the microwave background sky after subtraction of the dipole effect. *Bottom panel*: same but with the emission from the Milky Way masked out.

isotropic and homogeneous on these scales to this accuracy, 1 part in 100 000, at a time a few hundred thousand years after the Big Bang. This is a very strong statement about the smoothness of the universe and the degree to which it satisfies the cosmological principle.

To connect the COBE observations with the universe of galaxies today, we have to bridge these observations in two ways. We have to decide how structure on these very large scales connects with structure on the smaller scales of galaxies and clusters of galaxies. And we have to work out how we expect structure to evolve with time over the billions of years that the universe has existed. These issues will be explored in later chapters.

It is also natural to ask how these fluctuations originated. Were they present at the Big Bang, waiting for the opportune moment to develop into galaxies and stars? Did the universe start off even smoother, with the fluctuations appearing and growing at a later stage in its evolution? Or perhaps the universe did not start off from a smooth state at all but evolved into the relatively smooth state we see in the microwave background radiation? Cosmologists can only give answers to these questions if they are willing to extrapolate beyond what is reliably known about physics. There is no consensus yet and we are faced with a menu of speculations. In the course of this book I shall take you through this menu and explain what we really know about the origin of structure in the universe and what the prospects are for improving our state of knowledge.

But already we can conclude with some confidence that we are not in a special place and that the universe we live in has evolved from an earlier state extremely close to homogeneity and isotropy. This is a very deep fact about the universe.

Chapter 3

# An expanding universe

*The observations have disclosed the remarkable fact that in (the galaxies') spectra there is a displacement towards the red corresponding to a receding velocity increasing with distance, and, so far as the determinations of the distances are reliable, proportional with it. If the velocity is proportional to the distance, then not only the distance of any nebula from us is increasing, but all the mutual distances between any two of them are increasing at the same rate. Our own galaxy system is only one of a great many, and observations made from any of the others would show exactly the same thing: all systems are receding, not from any particular centre, but from each other: the whole system of galaxies is expanding.*

Willem de Sitter, *Kosmos*, 1932

In this chapter we explore what at the beginning of the twentieth century would have seemed a very surprising fact about the universe, that it is expanding. The third of our nine numbers of the cosmos will be the current rate at which the universe is expanding, which is measured by what is known as the Hubble constant. Apart from the wanderings of the planets and the occasional comet or supernova, the night sky does not seem to change, so it is natural that through most of human history philosophers have assumed a static universe, unchanging with time. Up to and including Einstein's landmark paper of 1917, almost all models proposed for the universe had been static. In Einstein's static model the tendency of gravity to make the matter in the universe fall together was balanced by a new force, the cosmological repulsion. Unfortunately this model was soon seen to be unstable. If you lit a match in this universe it would immediately start expanding or contracting.

## Problems with a static universe

When challenged by Richard Bentley in 1692 to explain what would happen to stars spread through the universe under the influence of the newly postulated universal gravitational force, Newton had come up with the argument that in an infinite universe a star would not know which way to fall so would stay still. Without the cosmological repulsion, this ambiguity disappears in Einstein's general relativity.

However, the infinite, static, infinitely old universe had other problems. Edmund Halley, a contemporary and friend of Newton's, pointed out that if you added up the light from all the stars in an infinite universe, the result would be infinite. Instead of being dark at night, the universe would be infinitely bright. More careful consideration leads to the conclusion that the sky should be as bright as the surface of the sun in every direction, because a straight line in any direction will eventually hit the surface of a star.

This paradox was restated several times, most forcefully by Wilhelm Olbers in the early part of the nineteenth century, and in modern times has been known as Olbers' paradox. However, the paradox had essentially been resolved by the poet and science writer Edgar Allan Poe in his treatise *Eureka* in 1848. He pointed out that if the universe is of finite age then the sum of the light from all the stars is finite. Most lines of sight reach back in time to the start of the universe before they hit a star. Poe also saw rather clearly the concept of a **horizon**, that if the universe is of finite age, we can not see out further than light can travel in that time. The whole story has been excellently recounted in Edward Harrison's 1987 book *Darkness at night*.

## Hubble's great discovery

In 1929 Edwin Hubble made one of the most dramatic scientific announcements of the century. The universe is expanding. The galaxies are moving away from us and the further away they are the faster they are receding. Progress on exploring the 'realm of the nebulae', as Hubble called it, had been very rapid. As recently as 1920 there had been a debate between two leading American astronomers, Harlow

Shapley and Heber Curtis, on the nature of the nebulae and the size of the Milky Way. Curtis had argued for the island universe picture, in which the spiral nebulae were distant galaxies like our own. Shapley believed that he could show that many of these spiral nebulae actually lay within the confines of the Milky Way system. His arguments appeared more systematic, but they were wrong on two counts. Firstly, because he was neglecting the effects of interstellar dust he was greatly overestimating the size of our Galaxy. And secondly, his measures of the distances of the nebulae were seriously underestimated. Better distances were derived for several galaxies by Hubble using Cepheid variable stars during the 1920s. Henrietta Leavitt of the Harvard Observatory had discovered in 1908 that there is a linear relation between the luminosity of Cepheid variable stars and their period of variation. Cepheids are massive stars which become unstable during a phase of their evolution and undergo regular pulsations. The star pulsates in and out and its brightness changes in a characteristic way. Leavitt's discovery, made by studying the variable stars in the Magellanic Clouds, was that the more luminous the star, the slower its pulsations. So by measuring the period of variation we can determine the luminosity, or total energy output, of the star. From the apparent brightness of the star (how much energy we receive at earth per second per unit area) we can then deduce the distance of the star using the inverse square law for brightness. By 1925 these improved distances to other galaxies based on Cepheid variable stars had swung the debate decisively in favour of the island universe picture. Meanwhile, Vesto Slipher had been measuring the velocities of the spiral nebulae with the Lowell 24-inch refractor, starting with the Andromeda nebula in 1913. By 1915 it was clear that the majority of the velocities Slipher was finding were positive; that is, the galaxies were moving away from us. By 1929 Hubble had velocities for 46 extragalactic nebulae, some of the more distant ones measured by his assistant Milton Humason using the new Mount Wilson 100-inch telescope. Hubble and Humason both had exotic backgrounds. Edwin Hubble was a good enough amateur boxer to have been offered the chance to turn professional. Instead he studied law before finding his true vocation in astronomy. Milton Humason left school at 14 and became a mule-driver during the construction of the Mount Wilson Observatory, before becoming janitor there, then telescope night assistant, and

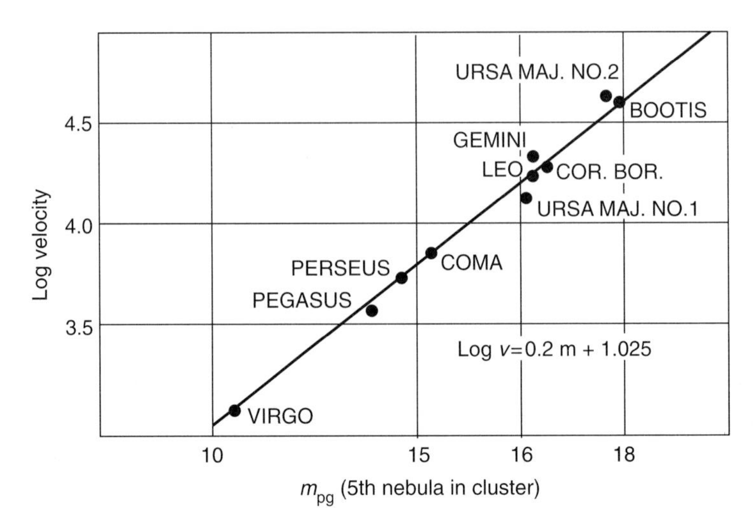


**Fig. 3.1** The Hubble law: the velocity–distance relation for galaxy clusters published by Edwin Hubble and Milton Humason in 1931. The horizontal axis is actually the photographic magnitude of the fifth brightest galaxy in the cluster. If these galaxies all have the same luminosity, the magnitude would be related to distance by $m_{pg}$ = 5 log (distance) + constant.

finally, having demonstrated his skill as an observer, an astronomer. Of the 46 galaxies for which Hubble now had redshifts, he had accurate distances for 24. This was sufficient for him to show that there was a clear increase of velocity with distance, the famous Hubble velocity–distance law.

In his 1929 paper, Hubble did not make much comment on the theoretical interpretation of this law. The last, rather convoluted paragraph of the paper remarked:

> *The outstanding feature, however, is the possibility that the velocity-distance relation may represent the de Sitter effect, and hence that numerical data may be introduced into discussions of the general curvature of space. In the de Sitter cosmology, displacement of the spectra arise from two sources, an apparent slowing down of atomic vibrations and a general tendency of material particles to scatter. The latter involves an acceleration and hence introduces the element of time. The relative importance of these two effects should determine the form of the relation between distances and velocities; and in this connection it may be emphasized that the linear relation found in the present discussion is a first approximation representing a restricted range in distance.*
>
> *Proceedings of the National Academy of Science*, 1929

The theorists of the time were still debating whether the universe obeyed Einstein's static model, in which no redshift is seen, or a model by Willem de Sitter in which the universe is expanding, and distant objects are redshifted, but the universe contains no matter. As can be seen from Hubble's comment, the interpretation of the redshift in the de Sitter models is ambiguous and Hubble is not making the claim here that the universe is expanding. Few astronomers were aware of the more general expanding universe solutions found by Georges Lemaître and, especially, Alexander Friedmann, who had found in 1922–24 a complete set of solutions for Einstein's equations for the cosmological problem. Following Hubble's paper it was quickly realized that the expanding universe solutions were the ones that were needed to understand his observations. Hubble himself was more cautious. In his 1936 book *The realm of the nebulae* he wrote, discussing the Doppler shifts seen in the spectra of stars (shifts in the wavelengths of lines in the spectra of the stars which are interpreted as due to their motion towards us or away from us):

> *The same interpretation is frequently applied to the red-shifts in nebular spectra and has led to the term 'velocity-distance' relation for the observed relation between the red-shifts and apparent faintness. On this assumption, the nebulae are supposed to be rushing away from our region of space, with velocities that increase directly with distance.*

The sceptical tone of these comments suggests that Hubble had by no means accepted the expanding universe interpretation of the redshift–distance relation. So perhaps Hubble is rather lucky to be credited with discovering the expansion of the universe. Between Hubble's day and our own, dissident scientists have repeatedly tried to invent other explanations of the cosmological redshift, but without any convincing success. One idea that has surfaced several times is the 'tired-light' hypothesis, according to which photons of light simply lose energy during their passage across the universe and thereby shift to lower frequencies or longer (redder) wavelengths. Combined with the normal inverse square law dimming of brightness of a radiation source with distance, the tired-light hypothesis would seem to be able to explain what we see without having an expanding universe. Most cosmologists are not very interested in this explanation, though, because it does not have a physical basis. According to general relativity and quantum theory, a photon of light does not become 'tired' as it

**Fig. 3.2** Willem de Sitter.

travels. Anyway, the recession of the galaxies causes an extra dimming effect in addition to the inverse square law effect of distance and this has now been identified in careful analysis of distant giant elliptical galaxies.

It is sometimes said by popularizers of science that you can think of the expansion of the universe as a stretching of space rather than a real motion of the galaxies away from us. Although the general theory of relativity does permit us to describe phenomena from any frame of reference, this expansion is still a pretty real phenomenon. If we set up an experiment in which we measure the distance to a galaxy by sending a radar signal to the galaxy that is reflected back to us (not a very practical possibility given the millions of years that would be needed for this experiment) and then carried out the same experiment at a later time, we would find that the distance had increased. So the expansion is real enough.

## The third of the nine numbers of the cosmos, the Hubble constant

Hubble's law gives us the third of our nine numbers, the constant of proportionality in the velocity–distance law, Hubble's constant, denoted by $H_0$ ('H-zero'), so velocity = $H_0 \times$ distance. Astronomers measure this in units of kilometres per second per megaparsec. One megaparsec is a million parsecs and a parsec is the distance at which the radius of the earth's orbit subtends an angle of an arc second (i.e. a parallax of 1 arc second), so 1 megaparsec (Mpc for short) = 3.26 million light years = 30.9 million million million kilometres. Now kilometres and megaparsecs are both units of distance, so the true dimensions of $H_0$ are 1/time. We call the time corresponding to $1/H_0$ the Hubble time, $\tau_0$. It measures the rate at which the universe is expanding today, the time for the universe to double in size expanding at its present rate. If the galaxies moved apart with no forces acting, then this would be the same as the age of the universe. If the only force acting is gravity, which slows down the expansion of the universe, then the age of the universe will be less than the Hubble time.

When Hubble announced his velocity–distance law in 1929, the value of the constant of proportionality which he found was 500 km $s^{-1}$ $Mpc^{-1}$. This implied a Hubble time of 2 billion years. This already seemed much too short compared with the estimated age of the Milky Way, which James Jeans had estimated to be 100 billion years by dynamical arguments (the latter was in fact a gross overestimate). In 1931 Hubble and Humason published a much improved velocity–distance diagram, which reached much deeper into the universe. Their value of the Hubble constant increased slightly to 550 km $s^{-1}$ $Mpc^{-1}$. In 1935 the British geologist Arthur Holmes announced an age for the earth derived from radioactive dating of 3.6 billion years, which he later increased to 4.3 billion years, not far off the correct value of 4.5 billion years. This was clearly inconsistent with the age of the universe implied by Hubble's law. We will return to this issue of the age of the universe in the next chapter.

## Sorting out the value of $H_0$ and the steady state cosmology

It was to take over 20 years to sort out this discrepancy. In the meantime, one response, by Herman Bondi, Thomas Gold, and Fred Hoyle in 1948, was to develop a new model of the universe in which the Hubble time had nothing to do with the age of the universe. This was the **steady state model**, in which the universe is postulated not only to be homogeneous and isotropic but also to present an appearance on the large scale which is unchanging with time. In order to maintain the average density of the universe at a constant value in the face of the expansion, the model's proponents had to postulate that matter is created continuously throughout the universe. At the time most cosmologists found the introduction of this additional 'creation field' into the basic equations of matter unsatisfactory and the model did not find many supporters outside the UK. Ironically today, although the steady state theory is long dead, the practice of adding additional fields, representing hypothetical forces, to the basic equations is now routine and respectable, at least in studies of the very early universe. The steady state model ran into difficulties in the early 1960s when increasing evidence of evolution of different populations of galaxies over cosmological times began to be found. The discovery of the microwave background radiation in 1965 was seen as confirmation of the rival Hot Big Bang model. By then the huge discrepancy between the Hubble time and the age of the universe had been, at least partially, resolved.

The first step was the realization by Walter Baade, working at Mount Wilson Observatory in 1952, that there are in fact two types of Cepheid variable star, satisfying different period luminosity relations. The Cepheids found by Hubble in nearby galaxies are massive stars in young clusters, Type I Cepheids, but the Cepheids in our Galaxy, which he was using to calibrate his distance estimates, were lower mass stars, Type II. Correcting this immediately reduced the Hubble constant from 550 to 250 km $s^{-1}$ $Mpc^{-1}$. Allan Sandage, hired by Hubble to continue his work on the distance scale at the Mount Palomar Observatory, found a further discrepancy in 1956. Sandage found that objects which Hubble had thought were the brightest stars in galaxies turned out to be huge hot gas clouds known as 'HII regions'. This brought the Hubble constant down to 75 km $s^{-1}$ $Mpc^{-1}$. From 1956 to the present, estimates of the Hubble constant have almost always been in the range 50–100 km $s^{-1}$ $Mpc^{-1}$, so we can see Sandage's 1956 paper as a landmark.

The exact value of the Hubble constant has, however, been a source of constant controversy during the past 40 years. I have described the history of attempts to measure distances and ages in the universe in my book *The cosmological distance ladder*, which takes the story up to 1983. At that time the controversy was seen as a debate between Sandage and his Swiss co-worker, Gustave Tammann, who favoured a value of 50 km $s^{-1}$ $Mpc^{-1}$, and the French astronomer Gerard de Vaucouleurs, who favoured a value of 100 for $H_0$. However, both camps overestimated the accuracy of their values. Much of their disagreement was caused by using very different distances for the very nearest galaxies in the Local Group (p. 23). As new methods were developed by younger astronomers a wide range of values for $H_0$ continued to be found. By 1990 there were still debates between advocates of 50 km $s^{-1}$ $Mpc^{-1}$ and those who favoured 90 or so. Only at the lower end of the range could consistency be found between the age of the universe and the age of the oldest stars.

## The Hipparcos and Hubble Space Telescope missions

In the 1990s two space missions began to dominate the debate over the Hubble constant. Both had disastrous beginnings. The Hipparcos mission, launched in 1989 by the European Space Agency, set out to monitor the positions of hundreds of thousands of stars to unprecedented accuracy. This was to place the local distance scale on a firm footing, but it also had the consequence (as we shall see in the next chapter) of revising the estimates of the ages of the oldest stars. When Hipparcos was launched, the final stage failed to ignite correctly and the satellite was not placed in its correct circular orbit. Instead it ended up in a highly elongated orbit. However, by redesigning the observing programme of the mission, the Hipparcos scientists managed to recover almost the full scientific performance planned. Results from this mission will be discussed in Chapter 4.

The Hubble Space Telescope, launched by NASA in 1990, had as one of its key programmes the goal of measuring distances to much more distant galaxies than was possible with ground-based telescopes and thereby of resolving the controversy over the Hubble constant. It was soon realized that the mirror had been polished to the wrong shape and the images seen by the telescope suffered horrible distortion. The error resulted from a trivial mistake, that a humble washer in the eyepiece used to monitor the polishing had been omitted. At first the distortion in the mirror shape threatened to make the goal of measuring galaxy distances unattainable, but the repair mission of 1993 provided a fix which has allowed the Hubble Space Telescope to achieve almost all its goals. Cepheid variable stars have now been studied in scores of galaxies at distances out to 50 million light years and already the range of disagreement over the Hubble constant has been sharply reduced. Hubble had used Cepheids in the

**Fig. 3.3** The Hubble Space Telescope.

1920s to establish that the spiral nebulae lie outside the Milky Way, and they have been at the heart of galaxy distance measurements ever since. The tremendous improvement in resolution that the Hubble Space Telescope offers due to being above the distorting effect of the earth's atmosphere allows much better discrimination of individual stars in external galaxies, so that even very faint Cepheids can be found and their brightness monitored. Wendy Freedman, of the Carnegie Observatories, Pasadena, has summarized her results and those of her colleagues with the Hubble Space Telescope and gives a combined value for the Hubble constant from all the galaxies observed as $H_0 = 75$ km s$^{-1}$ Mpc$^{-1}$, with an uncertainty of $\pm 15$.

## Supernovae and other routes to $H_0$

A second powerful distance indicator has begun to show tremendous potential to reach out to vast distances: the supernova or exploding star. Supernovae, where a star brightens by a factor of a million to a billion over a period of a few days, have been seen in the sky since ancient times, though only in the 1930s did astronomers begin to have an inkling of what was going on. The ancient Chinese astronomers, who watched the sky carefully each night for changes and omens, recorded several examples, and the Renaissance astronomers Tycho

Brahe and Johannes Kepler each identified and studied new supernovae. There is no clear case of a supernova visible to the naked eye since Kepler's supernova of 1604, though Supernova 1987A in the Large Magellanic Cloud, a small satellite galaxy of our own Galaxy, came close. It was the Swiss-American astronomer Fritz Zwicky who, in the 1930s, coined the term supernova (previously they had been called new stars) and suggested, partially correctly, that they were due to the explosion of massive stars.

There are two main types of supernovae: Type I (more particularly Ia, as there are some anomalous supernova types with a different explanation) are the more luminous and are caused when large amounts of gas are dumped on a white dwarf by its companion in a binary system. White dwarfs are very compact, dead stars, in which no nuclear reactions are taking place and the star is held up against gravity by the pressure of the electrons in the star being crushed together. Because they have a maximum stable mass of about 1.4 times the mass of the sun, if gas is added to the star and takes the mass over this limit, the star explodes. The second type of supernova, Type II, is caused when a star of greater than about eight times the mass of the sun comes to the end of its nuclear-burning life, when the core of the star is made of iron. At this point no further energy generation by nuclear fusion can occur because, as we have seen, when iron is fused to make heavier elements, energy is absorbed rather than released. The core of the star collapses to make a neutron star, an even more compact object than a white dwarf, or a black hole (if the core mass is greater than twice the mass of the sun), and the rest of the star is ejected in a spectacular explosion. The star brightens by a factor of a million in a few days, then gradually declines in brightness over the next months and years. For comparison, a white dwarf of the mass of the sun would be about the size of the moon, a neutron star would be about the size of Greater London or Los Angeles, and a black hole of this mass would have a horizon about a mile across.

Both types of supernova have been used to estimate distances, by carefully monitoring the rate of expansion of the ejected material and the changes of colour and hence brightness per unit surface area. The expansion rate gives the linear size of the envelope and the brightness per unit area, when combined with the observed total brightness, gives the angular radius, so the distance can be found. This method of

**Fig. 3.4** A supernova seen in 1972 in the outer parts of the galaxy NGC5253.

estimating distance is known as the Baade method, after Walter Baade who used it for variable stars. For Type Ia, the luminosity at maximum light is roughly the same for all supernova events so the brightness of a supernova at maximum light seen at the earth is simply inversely proportional to its distance. Since supernovae can now be detected at immense distances, they can be used to measure not only the Hubble constant but other cosmological parameters, as we shall see in Chapter 8. David Branch, of the University of Oklahoma, has reviewed many different groups' analyses of Type Ia supernova data and concludes that the best estimate for the Hubble constant from this method is $H_0 = 60$ km s$^{-1}$ Mpc$^{-1}$, with an uncertainty of ±10. Brian Schmidt, at Harvard, and his collaborators found $H_0 = 73$ km s$^{-1}$ Mpc$^{-1}$, with an uncertainty of ±13, for Type II supernovae.

In the past decade two new distance methods have emerged which also have the capability to reach out to very great distances. The first is based on the curious phenomenon of the **gravitational lens**. Einstein's general theory of relativity predicts that light is bent round any mass. This means that masses act like a lens, focusing the light from a source behind the lensing mass. The result is not a perfect lens and in

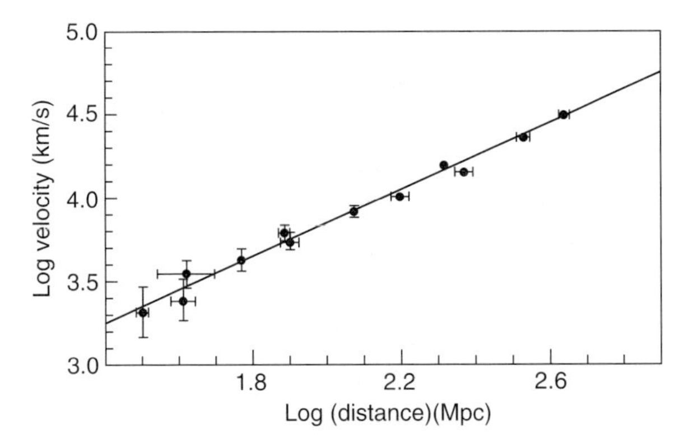


**Fig. 3.5** The Hubble diagram for Type Ia supernova. This analysis, by Adam Riess of the University of California, Berkeley, and colleagues, yielded a Hubble constant of 67 ± 7 km $s^{-1}$ $Mpc^{-1}$. From *Astrophysical Journal*, **438**, L17 (1995).

general several images are seen. If the lens and the source are perfectly aligned a point source of light is imaged into a ring the radius of which depends on the mass of the lens. If the alignment is not quite perfect the ring breaks up first into two circular arcs and then into two or more images seen on either side of the lensing object. Strictly, there have to be an odd number of images, but generally one of these is too faint to be seen. Gravitational lensing has now been observed on many scales.

Three major international experiments are looking for low-mass stars in the halo of our Galaxy via their lensing effects. The idea is to look for brightness variations of distant background stars as the lensing star moves across the line of sight. Every night each team monitors the brightness of around a million stars, selected to be either in an external galaxy like the Large Magellanic Cloud or in the distant central bulge of our Galaxy. If a dark star crosses the line of sight to the background star, the gravitational lens effect changes the brightness of the background star in a very characteristic way, first brightening it and then dimming it again over a time-scale of weeks or months. The breaking up of the stellar image into multiple sources, arcs or a ring, can not be seen because it is happening on too small an angular scale and this is known as **microlensing**. The time-scale of the brightening and dimming gives the mass of the lensing star or object. Dozens of lensing events have now been seen and the masses of the lensing objects are generally between a tenth of a solar mass and half a solar mass. The programmes were set up to look for brown dwarfs and Jupiter-mass objects, but none of these has definitely been seen. The detected objects are probably either white dwarf stars in our Galaxy or normal stars in the Magellanic Clouds, but they could conceivably be low-mass black holes in the halo of our Galaxy.

On a larger scale, multiple images and rings due to very distant galaxies and quasars have been found at optical and radio wavelengths due to lensing by intervening galaxies. In fact we now have to be suspicious any time that we come across a galaxy or quasar that seems exceptionally luminous, because in several cases in recent years such objects have turned out to be significantly amplified by lensing. When lensing by a single galaxy is involved the scale over which we see amplification and distortion of the image is of the order of 1 or 2 arc seconds, so the powerful resolution of the Hubble Space Telescope is an essential aid in mapping these. On a still larger scale multiple arcs are often seen in rich clusters of galaxies due to lensing of distant background galaxies by the whole cluster. Because lensing is affected by the whole mass of the lens, mapping and analysing the lensed image can be an effective way to probe dark matter within the lensing object. We will return to this in Chapter 6 when we consider the role of dark matter in the universe.

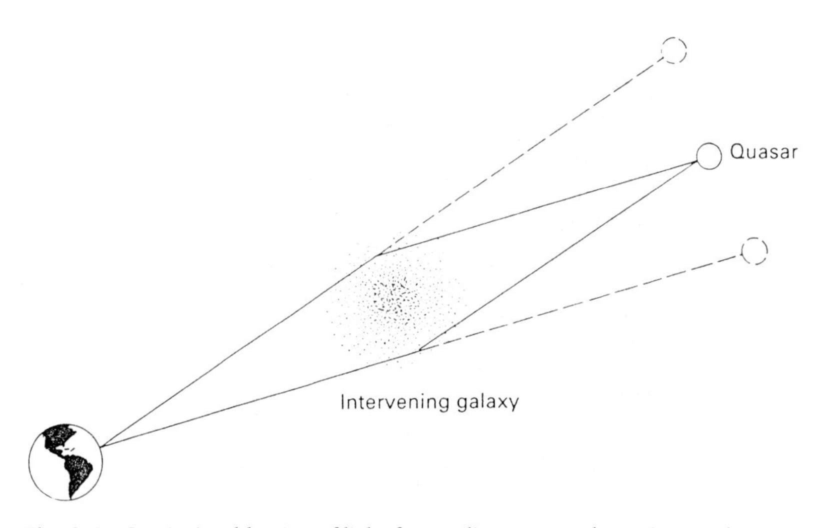


**Fig. 3.6** Gravitational lensing of light from a distant quasar by an intervening galaxy. The light is bent slightly as it passes around the galaxy, resulting in two or more images of the quasar.

For determination of distance, however, it is the multiple images that are of interest. If the background source is varying its light output with time, as often happens with quasars at radio wavelengths for example, then the multiple images will appear out of phase because the time for the light to travel by the two different routes will be different. In one well-studied system, known as 0957+561, there are two bright lensed images of a variable quasar and the time delay between the two signals is about 250 days. If we can work out the mass of the lensing galaxy, we can use this time delay to deduce the distance of the lens and the source, because we know the difference in the two lensing paths (250 light days). Emilio Falco, of Harvard, and collaborators have given a recent estimate for the Hubble constant of 62 ± 7 km $s^{-1}$ $Mpc^{-1}$, using the gravitational lens time delay in 0957+561.

A second new distance method is based on an effect discovered in 1970 by two Russian scientists, Rashid Sunyaev and Yakov Zeldovich. Rich clusters of galaxies contain very hot gas at a temperature of a hundred million degrees, which was discovered with the Uhuru satellite in the same year through its intense X-ray emission. The origin of this gas is not entirely clear but it was probably stripped out or blown out of the galaxies in the clusters. The gas is known to contain iron, detected through a characteristic X-ray emission line, and iron can only have been made in stars. If we map the microwave

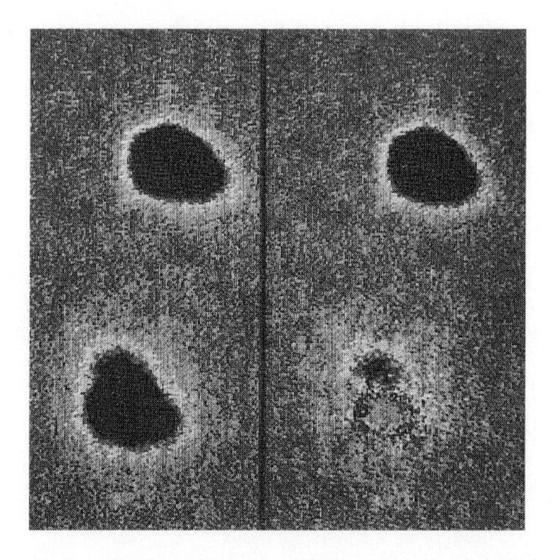

**Fig. 3.7** The gravitationally lensed quasar 0957 + 561. In the right-hand frame the upper image of the quasar has been carefully subtracted from the lower image to reveal the lensing galaxy. (Photo courtesy of Alan Stockton, University of Hawaii.)

background radiation around a cluster of galaxies we will notice a change in brightness of the background in the direction of the cluster. At wavelengths longward of 1 millimetre the cluster appears as a slightly darker patch, while at wavelengths shorter than 1 millimetre, the cluster appears slightly brighter. The reason is that the electrons moving around freely in the hot cluster gas interact with the photons of the microwave background and boost them to slightly higher energy, that is to shorter wavelengths. The amount of the dimming or brightening depends on the temperature of the gas, which can be determined from X-ray studies, on the density of electrons in the cluster, and on its size. The brightness of the X-ray emission from the cluster gas also depends on the density of electrons and the size of the cluster, but in a different way, so the linear size of the cluster gas cloud can be determined. Hence, from the angular extent of the cluster gas cloud the distance of the cluster can be determined, and so by measuring the recession velocity of the galaxies in the cluster we get the Hubble constant. Values in the range 40–70 km $s^{-1}$ $Mpc^{-1}$ have been found by this method. Mark Birkinshaw, of Bristol University, and collaborators have given an average value of 55 ± 17 km $s^{-1}$ $Mpc^{-1}$ from the Sunyaev–Zeldovich effect in two clusters and Steven Myers of the California Institute of Technology, and his colleagues have found 54 ± 14 km $s^{-1}$ $Mpc^{-1}$ for four clusters.

## Summary on the value of $H_0$ and future work

If we combine the values of the Hubble constant determined by Hubble Space Telescope studies of Cepheid variables stars, by supernovae, and by the gravitational lens time delay and Sunyaev–Zeldovich method (Table 3.1), giving equal weight to each determination, we find an average value of 65 km $s^{-1}$ $Mpc^{-1}$, with an uncertainty of ±8 km $s^{-1}$ $Mpc^{-1}$. This uncertainty, which scientists call the **standard error**, has the meaning that 95% of the time the measured quantity would be expected to lie within twice this value of the average value, which in this case means that the 95% range of uncertainty is 49 to 81.* The mean value is almost identical to the

* 65 is the straight average of the five values, giving them all equal weight. The standard error is then 8, giving a 95% uncertainty range of 49–81. If we take into account the accuracy claimed for each method, and weight the average by $1/(\text{uncertainty})^2$, we get 63 ± 5 km $s^{-1}$ $Mpc^{-1}$, so the 95% range becomes 53–73. In this book I shall use the first, more conservative, value.

Table 3.1 The Hubble constant

| | | |
|---|---|---|
| Ceheids in Virgo (M100) | $H_0 = 75 \pm 15$ | (W. Freedman *et al.* 1997) |
| Type Ia supernovae | $H_0 = 60 \pm 10$ | (D. Branch 1998) |
| Type II supernovae | $H_0 = 73 \pm 13$ | (B. Schmidt *et al.* 1994) |
| Gravitational lens time delay | $H_0 = 62 \pm 7$ | (E. E. Falco *et al.* 1997) |
| Sunyaev-Zeldovich effect | $H_0 = 54 \pm 14$ | (S. T. Myers *et al.* 1997) |
| **Average of these five:** | $H_0 = 65 \pm 8$ **km s$^{-1}$ Mpc$^{-1}$** | |

value I derived in my 1985 book *The cosmological distance ladder*, which was 66. It seems that after a 70-year search, with the past 20 years being a period of particularly vigorous controversy, we are finally close to a consensus on the Hubble constant.

There is still some work remaining to be done before we can be sure we have reached a definitive value. Each galaxy or cluster of galaxies suffers a small velocity deviation from the Hubble velocity-proportional-to-distance expansion because of the gravitational effects of other galaxies and clusters. These deviations are called 'peculiar' velocities, not because there is anything strange about them but because they are peculiar to (or particular to) the galaxy or cluster concerned. One problem is that the peculiar velocities of the Virgo cluster and other nearby clusters which can be calculated by using Cepheids need to be determined more accurately. This can be done using deep all-sky redshift surveys like the **IRAS PSCz** Galaxy Red Shift Survey, which my collaborators and I have been working on for many years, to map the three-dimensional galaxy distribution. We can then calculate what we expect the peculiar velocities of nearby galaxies or clusters of galaxies, due to the attraction of the other galaxies around them, to be. The IRAS PSCz Galaxy Red Shift Survey is based on the all-sky survey at infrared wavelengths made using the Infrared Astronomical Satellite (IRAS). I have told the story of this mission in my book *Ripples in the cosmos*. From this survey we have extracted a sample of 15 000 galaxies spread all round the sky and ranging out to 3000 million light years in distance. We have measured the redshifts of all these galaxies and are now using them to map the galaxy distribution, study large-scale structure, and estimate the amount of dark matter in the universe. The uncertainty in the peculiar velocities of galaxies whose distances have been measured using Cepheid variable stars represents the main uncertainty in determining the Hubble constant by this method.

We also need to have supernova distances derived using the full geometrical method I described above, the Baade method, for a large sample of supernovae reaching out to distances as great as possible. And we need many more examples of the gravitational time delay and Sunyaev–Zeldovich method, to be sure there are not systematic problems with the methods. We can not be very confident about an estimate of the Hubble constant based on a single system.

## Using the Hubble constant to obtain a dimensionless measure of the average density of the universe

As I mentioned at the beginning of this chapter, the Hubble constant has the dimensions of 1/time. In Chapter 2 I said that we would like to measure cosmological parameters in a dimensionless form if possible and it was unsatisfactory to have the density of matter in the universe in terrestrial units of kilograms per cubic metre. We can use the Hubble constant to define a dimensionless density parameter, $\Omega$. For the reader interested in the precise definition, it is

$$\Omega = 8\pi G\rho/3H^2,$$

where $G$ is the gravitational constant, $\rho$ is the density, and $H$ is the Hubble constant. Now $H$ is not really a constant, since the expansion of the universe is expected to be slowing down due to the effects of gravity, and so $H$ changes with cosmic time. We add the subscript zero when we want to denote the value of the Hubble parameter at the present cosmic epoch, so it becomes $H_0$. Clearly it is $H_0$ that we are measuring in all our different distance experiments today. The equation above defines the density parameter at an arbitrary epoch. When we need to emphasize that we are talking about the value of the density parameter today we write $\Omega_0$.

From the value for the density of baryonic matter we quoted in Chapter 2, and the value for the Hubble constant, $H_0 = 65$ km s$^{-1}$ Mpc$^{-1}$, we arrive at a value for the baryonic density parameter, $\Omega_b$, of 0.03, with an uncertainty of 20%, where we have added a subscript b to emphasize that we are talking about the density of baryonic matter (protons and neutrons). Why we need to do this will become clear later.

It might seem unsatisfactory to take a rather well-determined quantity, the average density in baryonic matter, and combine it with a not-so-well-determined number, the Hubble constant. It is true that much of the uncertainty in $\Omega_b$ comes from the uncertainty in the Hubble constant. However, it turns out that $\Omega$ has quite a fundamental importance in determining the dynamics of the universe. In the absence of other forces like the cosmological repulsion (see Chapter 8), the fate of the universe depends on whether the self-gravity of the matter in the universe is sufficient to halt the expansion and turn the expansion into a contraction. If that happened, the universe would collapse together again and end in a Big Crunch, with the contracting phase almost a mirror image of the expanding phase. Almost, but not quite. For even if a direction of time is hard to sense on the largest scale of the universe or on the smallest quantum levels, it is clearly present in our consciousness and in all those other irreversible processes described by the second law of thermodynamics. One way of stating this law is that in a closed system, disorder increases. Another is that heat does not flow from a colder body to a hotter. So the direction in which time is flowing in the universe is the one in which stars are radiating away their heat and, eventually, when they run out of energy, dying.

The density parameter determines the fate of the universe. If $\Omega > 1$, the density of the matter in the universe is sufficient for the expansion to be reversed and for the universe eventually to collapse together in a Big Crunch. If $\Omega < 1$, the expansion will continue indefinitely, with the universe gradually becoming ever colder. Eventually the galaxies move apart at constant speed, with no deceleration. For $\Omega = 1$, the so-called **critical density** case, the universe keeps on expanding but the rate of expansion gets ever slower and slower. The density parameter $\Omega$ can be thought of as the ratio of the density of the universe to the critical value. It can also be interpreted as the ratio of the gravitational energy of a volume of the universe to the kinetic energy, or energy of motion, of the same volume; so if $\Omega > 1$, gravity wins and the universe eventually recollapses, but if $\Omega < 1$, the expansion prevails.

If there were only baryonic matter in the universe, then because the density parameter for baryons is much less than unity, we could be sure that the universe will expand for ever. But, as we shall see, there are strong arguments for believing that there is more to the universe than meets the eye.

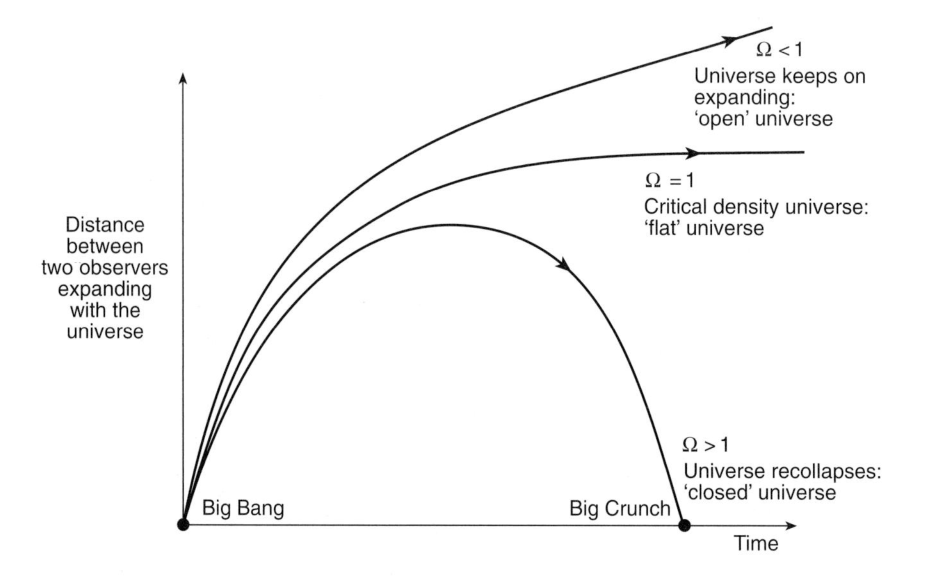


**Fig. 3.8** Three possible fates of the universe, depending on the value of the density parameter $\Omega$. For $\Omega > 1$, the universe recollapses to a Big Crunch. For $\Omega < 1$, the universe keeps on expanding forever. Between the two extremes is the $\Omega = 1$ universe, in which the expansion rate gets ever slower and slower.

Chapter 4

# A universe of finite age

*This world has persisted many a long year having once been set going in the appropriate motions. From these everything else follows.*

Lucretius, *On the nature of the universe*

The question of whether the universe has existed for ever has exercised the minds of scientists and philosophers from the earliest times. It is also the subject of myth and speculation in almost all cultures and religions. In Western Judeo-Christian culture we have the dominant myth of a Creation, which occurred at some particular moment of time, as recorded by the nomadic authors of 'Genesis' some time in the second millennium BC. For modern cosmologists the age of the universe is a purely scientific question and we have a range of techniques to estimate this age, which constitutes the fourth of the nine numbers of the cosmos. In the simplest cosmological models derived from Einstein's general theory of relativity, the age of the universe depends only on the Hubble constant and the density of the universe. From the time of Hubble to the present there have been worries that estimated values of the age of the universe, the Hubble constant, and the density of the universe are inconsistent with these simple cosmological models. This has led some cosmologists to favour models which give longer ages, for example by the introduction of a new force in the equations governing the dynamics of the universe, the cosmological repulsion. Current estimates do not really require this step, as we shall see later in this chapter. The subject of the cosmological repulsion is, for the most part, postponed to Chapter 8.

Although we are rather confident now that we live in an evolving universe in which the expansion began 10–15 billion years ago, we can not really trace the history of the universe back to the instant when the expansion began. Our current theories can take us back only to a minute fraction of a second after the Big Bang, the so-called 'Planck' time (see below). We can not rule out the possibility that the universe existed for an infinite time before that instant, or that it has been through many cycles of expansion and contraction.

## Greek ideas about time

The Greek philosophers entertained a wide range of possibilities for the history of the universe. Aristotle summarized the views of (most of) his contemporaries as:

> *All thinkers agree that this world has a beginning, but some maintain that having begun, it is everlasting, others that it is perishable like any other formation of matter, and others again that it alternates, being at one time as it is now, and at another time changing and perishing, and that this process continues unremittingly.*

Among the pre-Socratics, the Pythagoreans identified time with the motion of the celestial sphere. The atomists also had a relational rather than an absolute notion of time and identified time with the motion of atoms (time is 'a mere appearance'—Democritus). Parmenides, on the other hand, denied the reality of succession or change: these are illusions generated by our mode of perception. He believed, therefore, that the universe had always existed. For Aristotle, however, time had an absolute reality, is everywhere alike simultaneously, is present even if motion is absent. But he did recognize that a prerequisite of time measurement is a periodic mechanism, the best being the revolution of the celestial sphere. The Stoics believed that there was an extramundane void surrounding the material world. This void coexists with the world, so time has to flow in the void and must be absolute. They had a cyclical cosmological theory in which the universe is created from fire, and at the end of each cycle is dissolved in the original fire. This coincides with the beginning of another cycle in which the events of the previous cycle are reconstructed in all their details and in the same order.

For over a millennium the writings of the Greeks were lost to European view, preserved for us only by the scholarship and dedication of the Arab philosophers. The Moorish conquest of Spain brought to Europe not only the writings of Aristotle, but also sophisticated commentaries on them by the likes of Avicenna (Ibn Sina, 980–1037) and Averroes (Ibn Rushd, 1126–98).

The medieval theologians began to realize some disturbing implications of these writings. Saint Augustine, Bishop of Hippo in North Africa, had raised the thorny problem of the infinite divine idleness. If there had been an infinite time before the Creation, how had the Creator passed that time? Augustine's solution was to postulate that time began with the universe, a solution that for different reasons is still attractive today.

The alternative view, that the universe has always existed, was advocated by Parmenides in the sixth century BC, but did not find many supporters in the West. The attempt of the steady state theory of Bondi, Gold, and Hoyle, in 1948, to revive this view today appears as an interesting historical footnote. In 1964, however, when I began postgraduate work in cosmology, the steady state theory was a potent rival to the general relativistic models. Only with the discovery of the microwave background radiation and the growing evidence for evolution in the universe did the model gradually die. However, the Parmenidean view has revived again recently in the form of ideas about the very early universe like 'chaotic inflation', in which the universe existed for an infinite time as a fluctuating vacuum, until our region of it started to expand exponentially in an inflationary phase before commencing the more restrained expansion we see today.

## Growing realization of the antiquity of the universe

In the seventeenth and eighteenth centuries the date of the Creation was still accepted by most as the 4004 BC of Archbishop Ussher, arrived at by counting the generations listed in the Bible. However, the development of the science of geology by James Hutton (1726–97) and Charles Lyell (1797–1875), and the growing interest in fossils as a record of past species on earth, began to dent this picture. By the time of Darwin's *Origin of species* in 1859 it was clear that hundreds of millions of years were needed to account for the fossil record. Physical arguments were also resulting in much longer estimates for the age of the earth and the sun. In 1854 Helmholtz estimated, from considerations of the sun's gravitational energy, that it had been in existence for 20–30 million years and would last another 10 million years. Kelvin believed in 1862 that he had shown from cooling arguments that the habitable age of the earth could not be more than 200 million years.

The interesting argument by Edgar Allan Poe in his *Eureka* in 1848 has already been mentioned (p. 40). He linked the darkness of the night sky to the postulate that the stars have only radiated for a finite time.

In 1928 James Jeans estimated from dynamical arguments that the age of stellar systems like globular clusters and elliptical galaxies must be of an order of a hundred billion (100 thousand million) years, a huge overestimate. His argument was based on the observation that these systems appeared to be in statistical equilibrium, with the stars of the system all sharing roughly equal energy of motion. He assumed that this equilibrium was established by binary encounters between pairs of stars, but we now know that a much faster global process is at work.

Einstein's first model for the universe in 1917, his static model with gravity balanced by a cosmological repulsion, had an infinite age. Only when Lemaître and Friedmann began to look at expanding universe models did the possibility of a universe of finite age emerge. Hubble's discovery of the expansion of the universe appeared to support such models but seemed to give an age for the universe shorter than the age of the earth. Not till Allan Sandage's landmark 1956 paper (see p. 46) did a more reasonable estimate of the age of the universe emerge.

## The expanding general relativistic cosmological models

Within Einstein's general theory of relativity the expanding cosmological models satisfying the cosmological principle of homogeneity and isotropy can still have either finite or infinite age, if we allow the possibility of a cosmological repulsion. There are two kinds of universe with infinite age. The first starts from a static state at an infinite time in the past and expands away from it under the action of cosmological repulsion. These are known as the Eddington–Lemaître models. In the second type, the universe starts off collapsing inwards but the collapse is reversed by the cosmological repulsion and turned into an expansion—the so-called 'bounce' models. Both types of model can be ruled out because we know the average density of the universe is at least equal to the known density of baryonic matter. For the bounce model this would imply that the change from infall to expansion would have had to happen relatively recently and we would not be able to see highly redshifted objects like quasars. In the Eddington–Lemaître models the infinite age allows us to see right round the universe and in a model with density as high as that observed we would be starting to see the same objects a second time long before the redshift of the highest quasars was reached (see later). So if we are strictly within the framework of general relativity then we do not expect the universe to be of infinite age.

The simplest model of all is one in which the universe expands from the Big Bang and just keeps on expanding for ever. In such a model time is asymmetric, finite towards the past, infinite towards the future. This is the type of model which present observations favour. Aristotle, incidentally, thought such a model was ridiculous. He thought time should either be finite in both directions or infinite in both directions.

## The Planck time

Another model which could be relevant to the observed universe is the 'oscillating' model, in which the universe expands from a Big Bang; but eventually gravity halts the expansion and the universe recollapses to a Big Crunch (this happens to satisfy Aristotle's requirement that time is either finite in both directions or infinite in both directions). Mathematically the solution permits the universe to pass right through the Crunch and reemerge in a new expansion phase, hence the 'oscillation'. In practice we do not know how the universe could pass through a phase of infinite density. In physics a situation where some quantity goes infinite is known as a 'singularity'. General relativity would have broken down as a description of the universe well before this singularity was reached. We can characterize the point at which general relativity has to fail in terms of the **Planck time**. This is the moment immediately after the Big Bang, and there is another immediately before the Big Crunch, when general relativity has to break down and be replaced by some kind of quantum theory of gravity. Several ideas for such theories exist, but there is no consensus on the way forward. The most promising line of attack at the moment appears to be **superstring theory**, in which particles become tiny closed loops in a space of ten or more dimensions.

To see how we arrive at the magnitude of the Planck time, we have to compare two length scales, the radius of the universe and its

'Compton radius'. For a particle of a given mass (in this case the mass of the whole universe), the Compton radius defines the quantum uncertainty in position of that mass (according to quantum theory there is an uncertainty in precisely where any particle is at any time). The result of requiring these two length scales for the universe to be equal and dividing the resulting length by the speed of light is a time of $10^{-43}$ seconds, an unimaginably small instant of time.

Prior to the Planck time we do not really know what happened to the universe. In models which face a Big Crunch we do not know what the fate of the universe will be when it is within one Planck time of the Crunch. Thus theoreticians, or for that matter theologians, are free to speculate about what happened during these phases. There are speculations, for example, that the universe existed for an infinite time prior to the Big Bang as a fluctuating void, before our region of it embarked on the expanding phase we now inhabit. However ingenious theoretical work on these models may be, the models can not become established unless there is some observational evidence to support them. It is hard to see how this can happen and it seems unlikely that much progress will be made during the next century. In fact, if I had to choose one area in which science will not make much progress during the next millennium, it would be the history of the universe before the Planck time.

In talking about the age of the universe we therefore have to set aside our ignorance of what happened before the Planck time. What we are calling the age of the universe is the age since the Planck time, the age of our expanding phase.

## Radioactive dating

While nineteenth-century geology pointed the way towards a very long age for the earth, it was twentieth-century physics, in the form of radioactive dating, which actually determined this age. Already in the 1930s radioactive techniques were beginning to point towards ages of several billion years, in conflict with Hubble's wild underestimates of the age of the expanding universe (see Chapter 3). By the 1960s, when the Apollo programme succeeded in bringing back rocks from the moon, the age of the earth was known to high precision, 4.55 billion years. The moon rocks turned out to have exactly the same age. It became clear that the whole solar system formed at about the same time. But is our sun young or old? How long did the Milky Way exist before the sun formed?

Radioactive dating methods can be applied to our Galaxy too. Many of the elements on earth, including all the radioactive ones, were synthesized in supernova explosions before the sun formed. The sun and planets were assembled from a cloud of gas and dust which had been polluted with the products of successive generations of supernovae. As radioactive elements decay they gradually transform into their 'daughter' elements. For example, the half-life for the decay of uranium-238 to lead-206 is 4.5 billion years (so every 4.5 billion years, half the uranium atoms decay). If the decay is rapid we find only the daughter element today. In other cases we find a mixture of the radioactive element and its daughter, but in different proportions to those that are laid down in supernovae explosions. From analysing these proportions we can get an age estimate. But to get an age for the Galaxy we have to have some idea about the history of the formation of the supermassive stars that become supernovae. If they all formed together at the formation of our Galaxy, then we get a rather direct estimate of age. However, if stars have formed continuously over a long period of time, so that the abundance of the daughter element reflects a proportion of old supernovae where a lot of radioactive decay has occurred and of relatively recent supernovae from which the radioactive elements have not decayed much, then we must allow for this. Currently radioactive dating of our Galaxy results in ages between 10 and 15 billion years, depending on the assumptions about the star formation history.

## The age of the oldest stars

There is a second powerful method for estimating the age of our Galaxy, which involves studying the oldest stars in the Galaxy. These are the stars in globular clusters, compact, spherical systems containing perhaps a million stars within a region only 10 light years in diameter (a volume which in the neighbourhood of the sun would only contain a handful of stars). Now a natural assumption is that all the stars in a globular cluster system formed at the same time. The more massive stars evolve quickly and die, the less massive ones have hardly changed at all. In between there are the stars whose lifetime is of the same order as the Galaxy, which are just now changing rapidly as they approach death. By recognizing which stars these are from their colour and luminosity we can get quite an accurate estimate of the age of the Galaxy. In practice we model the changes in the whole population of stars in a globular cluster and from their colours and luminosities get an accurate estimate of the age of the cluster. For many years this method was quoted as giving ages in the range 14–18 billion years and this posed immense headaches to cosmologists, who could not get the age of the universe much above 13 billion years without invoking additional forces like the cosmological constant. Recently an important breakthrough has come from the European Space Agency's Hipparcos satellite, which was launched in 1989 to measure the positions of tens of thousands of stars to very high accuracy. One of the by-products of this mission was the realization that we have been underestimating the distances of the globular clusters, and hence the luminosities of the stars in the cluster. This in turn means that the stars in globular clusters are not as old as we previously thought. In a very thorough recent study that takes the new Hipparcos data into account, Brian Chaboyer, of Steward Observatory, and his collaborators conclude that the age of the oldest stars in globular clusters is 11.5 ± 1.3 billion years, so that the 95% range of values (see p. 54) is 9–14 billion years.

One further method of estimating the age of our Galaxy is to study the luminosities of white dwarf stars, another kind of old star. These are formed when a star like the sun exhausts its nuclear fuels and dies, blowing off its outer layer in a planetary nebula event and leaving the dense, degenerate core behind to cool off over billions of years. How faint white dwarf stars can get tells us how long they have been cooling off and hence gives the age of the Galaxy. However, since white dwarf stars are seen mainly in the disc of the Galaxy, they give an age only for this component. It would be possible for the halo of the Galaxy, where globular clusters are found, to be older than the disc. The age derived from white dwarfs is in the range 9–11 billion years, which may support this point of view.

## Conclusions on the age of the universe

Putting these estimates together we can conclude that the age of our Galaxy is probably in the range 10–13 billion years. Now we believe that a galaxy like ours formed quite early in the history of the universe,

**Fig. 4.1** The globular cluster 47 Tucanae.

0.1–1 billion years after the Big Bang. So the fourth number of the cosmos, the age of the universe, $t_0$, is probably in the range 10–14 billion years.

Can we get any support for the idea of a universe of this age from studying other galaxies? For some nearby galaxies, like the Magellanic Clouds, we can estimate the ages of their stellar clusters. Quite a range of ages is found. Reassuringly, none are found older than the age we derived above. However, it is difficult to get any kind of precision age for any other galaxy. The colours of galaxies are consistent with a scenario in which galaxies follow a star formation history similar to that we deduce for our own Galaxy. So nothing is inconsistent with

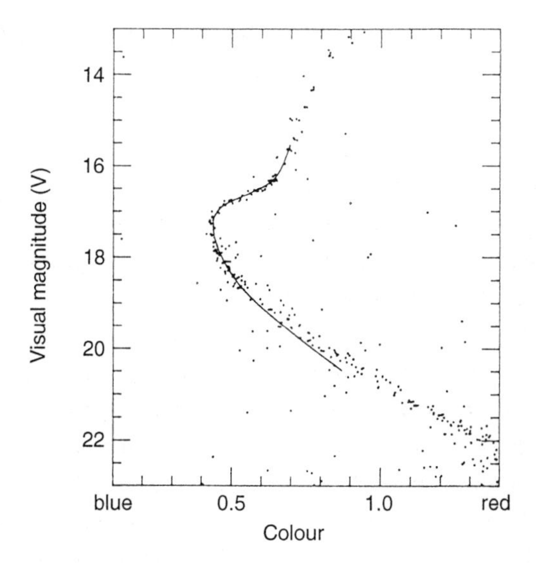


**Fig. 4.2** The colour–magnitude (Hertzsprung–Russell) diagram for the globular cluster NGC6752, studied by Alan Penny and Bob Dickinson. The observations are compared with the predictions of a theoretical model (solid line). If the distance of the cluster can be determined the age of the cluster can be deduced. The vertical axis is the visual magnitude (V) and the horizontal axis is the colour difference between the blue and the visual (green) bands (B – V).

**Fig. 4.3** The structure of our Galaxy, showing the disc, bulge, and halo of globular clusters.

our picture of a universe in which all galaxies start to form at the same time. But it would be nice to have some more precise ages of external galaxies. What we can say is that as we look out to greater distances, and so look back in time, the galaxies do seem to look more youthful, with more of them showing enhanced star formation activity and a much greater frequency of galaxy interactions and mergers. So, as expected, when we look back to earlier times, we appear to see the galaxies being assembled out of smaller pieces.

We now have two cosmic numbers with the dimensions either of time ($t_0$) or of 1/time ($H_0$). Given our aim of expressing all quantities in a form free of terrestrial units, we should multiply these together to get a dimensionless number, the ratio of the age of the universe to the Hubble or expansion time. If we do this using $H_0 = 65 \pm 8$ km s$^{-1}$ Mpc$^{-1}$ (so the Hubble time $\tau_0 = 1/H_0 = 15$ billion years, see p. 45) and $t_0 = 11.5 \pm 1.3$ billion years, where the uncertainties are as before chosen so that twice the uncertainty represents the 95% range in which we think the values could possibly lie (i.e. 49–81 km s$^{-1}$ Mpc$^{-1}$, 9–14 billion years), then the product of these becomes

$$\textit{age of the universe/Hubble time} = t_0/\tau_0 = H_0 t_0 = 0.80 \pm 0.13.$$

What do the simplest general relativistic cosmological models predict? For a universe of low density today ($\Omega_0$ small compared with 1), the expectation would be that $H_0 t_0$ would be close to 1. For a universe close to the critical density ($\Omega_0$ close to 1), we would expect $H_0 t_0 = 2/3$. The observed value lies between these values, but the full range of uncertainty (0.54–1.06) comfortably includes both extremes. So we may not yet be able to decide which of these possibilities is correct, but at least we do have two sensible possibilities. The revision to the ages of the globular clusters implied by the Hipparcos results is crucial to this statement. Since the time of Hubble we have almost always been in the situation that this product seemed to be greater than 1 so that additional forces like the cosmological constant had to be invoked. Today we are in the happy position of there no longer being (for the moment) an 'age of the universe' problem.

## Chapter 5

# The Hot Big Bang

*Ten billion years before now,*
*Brilliant, soaring in space and time*
*There was a ball of flame, solitary, eternal,*
*Our common father and our executioner.*
*It exploded, and every change began.*
*Even now the thin echo of this one reverse catastrophe*
*Resounds from the furthest reaches.*

Primo Levi, *In the beginning*

The discovery of the expansion of the universe by Hubble immediately posed the problem of what happens as we extrapolate the expansion back in time. As we run the universe backwards in time we would soon reach a phase where the galaxies all merge together. Further back still the stars would all be touching. And further still atoms would be crushed together.

The first to speculate on these questions was the Belgian Georges Lemaître. He started to apply thermodynamic considerations to the evolution of the universe and in 1931 came up with the concept of 'the primeval atom' from which the material of the universe was supposed to have fragmented (Chapter 1).

Also in the 1930s, the British theoretician Richard Tolman began to consider the possibility that the universe must contain radiation as well as matter and realized that the relative importance of radiation would increase as we look back towards the early universe. As the universe expands, the volume increases and the average density decreases inversely with the volume, that is inversely with the cube of the size of the universe. For the energy density of radiation there is an additional factor that the photons are redshifted to lower energy as the universe expands, their energy decreasing inversely with the size of the universe. Thus the energy density of the radiation decreases inversely with the fourth power of the size of the universe. So no matter how small a fraction of the total energy density of matter + radiation is in the form of radiation today, as we look back to the early universe there will eventually be an epoch when radiation was the dominant energy density. For the first time Tolman discussed the concept of the temperature of the universe and showed that this would increase into the past inversely proportionally to the size of the universe. The concept of the Hot Big Bang was beginning to emerge, in theory at least. However, other theorists continued to explore the idea that the universe had had a cold beginning and that stars and radiation emerged only later. Our fifth cosmic number will be the temperature of the universe today, as measured by the microwave background radiation.

## The Hot Big Bang and the microwave background radiation

The individual who did most to promote the concept of the Hot Big Bang universe was the Russian cosmologist George Gamow, who set out to explore in detail what such a universe would be like. His goal was to show that in a universe in which radiation dominates in its early phases, nuclear reactions could explain the origin of the elements. The discovery of the neutron by James Chadwick in 1932 allowed Gamow to start from a more realistic initial state than Lemaître, with a dense gas of neutrons, protons, and electrons, which, as we noted on p. 10, he called the 'ylem'. Although he was correct that nucleosynthesis processes take place during the hot phase of the Big Bang, in fact as we saw in Chapter 1 only the light elements are synthesized in any significant abundances. The Burbidges, Fowler, and Hoyle (see p. 11) showed in 1956 that elements from carbon upwards are made in stars. Their motivation was quite different. They hoped to show that *all* elements were made in stars so that the steady state theory would appear more plausible than a Big Bang model. Helium was the stumbling block for them and Hoyle was never able to show convincingly how helium could be made in sufficient quantities in a steady state universe. Back in 1949, Gamow's associates Alpher and Herman had predicted, in the framework of Gamow's goal to generate all the elements cosmologically, that the residual temperature of the background radiation today would be 5 degrees above the absolute zero of temperature, which is at –273 degrees Celsius (denoted 5 degrees Kelvin, or 5 K). At absolute zero the random motions of atoms in a gas or liquid, or of electrons in a solid, cease and there can be no lower temperature. For very low-temperature phenomena it is more convenient to use the Kelvin scale, which starts from absolute zero. However, Alpher and Herman do not seem to have made the connection between this temperature and microwave radiation.*

Soviet cosmologists, led by Yakov Zeldovich, were also pursuing similar lines of argument and they understood clearly that radiation with a temperature of 5 or 10 degrees Kelvin would be detectable as a microwave background radiation. They realized, too, that the key instrument for detecting this would be the microwave antenna at Bell Labs. They read the Bell Labs' technical journal to see what the capability of this telescope was. Had they read the reports more carefully they would have realized that there was a longstanding problem of excess and unexplained noise with the antenna, corresponding to an antenna temperature of about 3 degrees Kelvin. However, they misunderstood the different definitions of temperature used in these reports and concluded that there could be no cosmologically significant microwave background radiation. Zeldovich turned to consider cold Big Bang models, with no primordial radiation.

In 1963 two young radio-astronomers, Arno Penzias and Robert Wilson, who had recently moved to Bell Labs, decided to try to make some very accurate measurements of bright radio sources, particularly the supernova remnant known as Cassiopeia A, which would be valuable for calibration of other radio studies. They also wanted to measure the radio emission from the Milky Way at higher frequencies than had been attempted before, to study the physics of the radiation process. For both purposes the microwave antenna at Holmdel, built

* Nor, following the improved discussion of the physics of the early universe in 1953 by Alpher, Follin, and Hermann (see p. 10), did they recalculate their predicted temperature, or recalculate the abundances of the elements produced in the early universe. These calculations were not done until the work of Peebles in 1964 and Wagoner, Fowler, and Hoyle in 1966.

**Fig. 5.1** The Bell Labs' antenna with which the microwave background radiation was discovered.

for communications purposes, was ideal. First they had to deal with the well-known (at Bell Labs) problem of the excess noise of the antenna. Despite very intensive efforts they did not manage to eliminate this noise. When they heard that cosmologists were very interested in the idea of background radiation at microwave wavelengths, they realized what they had discovered and published quickly.

Over the months that they had been studying the microwave background radiation, Penzias and Wilson had found that it had the same intensity in whichever direction they looked. They also soon found that the temperature of the radiation was the same whatever wavelength they observed at. Over the next 25 years both the isotropy and spectrum of the microwave background radiation became a matter of intensive study. In Chapter 2 we saw that first the dipole anisotropy of the radiation due to our Galaxy's motion through space was discovered in the 1970s and then in 1992 the minute ripples in the background on scales of degrees were found by the COBE satellite team. COBE was in fact launched in 1989 and from data taken within minutes of starting observations John Mather, of Goddard Space Flight Center in Maryland, and his colleagues had found that the spectrum of the background radiation was a perfect black body spectrum (see p. 34) with temperature $T_0 = 2.728 \pm 0.004$ K. This type of spectrum was what was predicted for the relic radiation of the Hot Big Bang. The fifth of our cosmic numbers, the temperature of the microwave background, $T_0$, is easily the most accurately known.

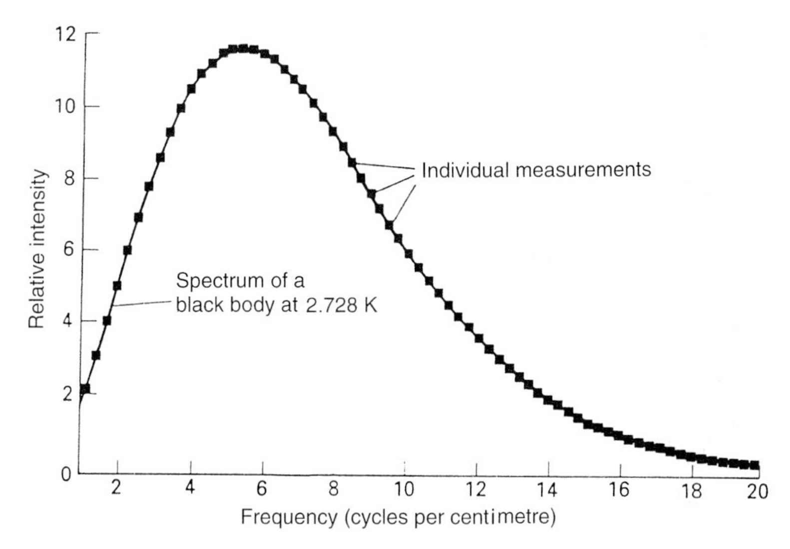


**Fig. 5.2** The spectrum of the microwave background radiation measured by John Mather and colleagues using the Cosmic Background Explorer (COBE) satellite in 1989. There is an extremely good fit of the observed points (solid squares) to a Planck black body curve.

The whole universe is bathed in this radiation at a temperature of 2.728 K, so apart from regions shielded from the radiation this represents the minimum temperature that matter in the universe can have. In fact this aspect of the background radiation, that everything is bathed in it, nearly led to its discovery in 1941. The Canadian astronomer A. McKellar noticed that a species of interstellar molecule, cyanogen, caused an absorption line in the spectrum of a star which implied a temperature of this order, but he did not appreciate the significance of this observation.

As we go back in time the radiation temperature satisfies the law, discovered by Tolman, that the temperature increases inversely proportionally as the size of the universe gets smaller. When the universe was ten times smaller, the temperature was 27.28 K and so on. We can imagine running the universe backwards in time and seeing what happens. The average spacing between galaxies today is a few million light years and the typical size of galaxies is about one hundredth of this, so when the universe was smaller than at present by a factor of 100, the galaxies (if they still existed unchanged) would all be touching. We can presume that the moment when galaxies took on a separate identity and started to form came after this. It is also reasonable to suppose that the first stars did not form till after this

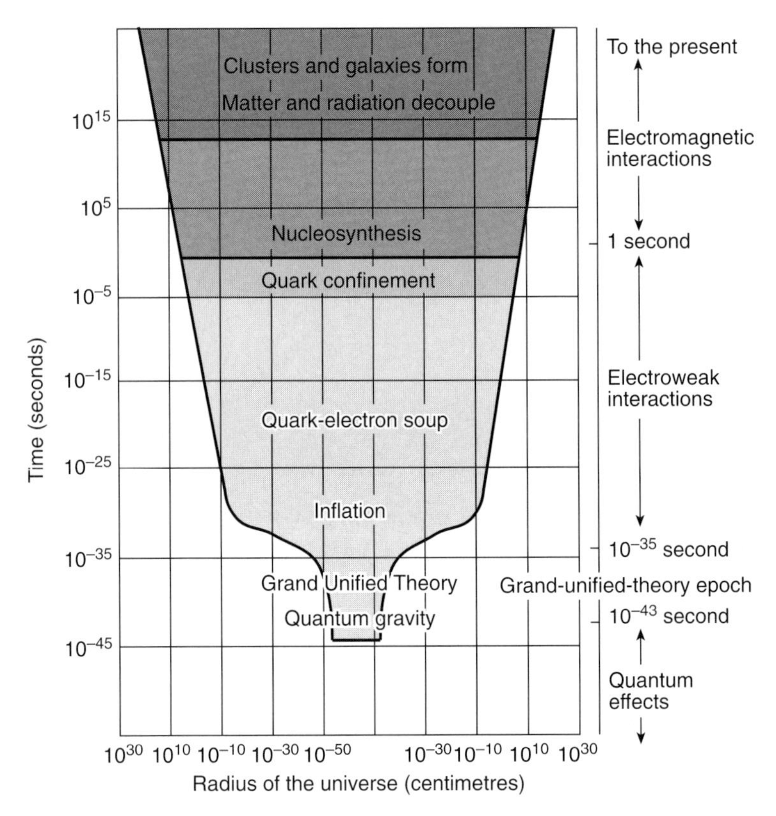


**Fig. 5.3** Schematic picture of the evolution of the universe. In the $10^{-43}$ seconds following the Big Bang, quantum effects dominate and the four fundamental forces (electromagnetism, weak and strong nuclear forces, and gravity) are believed to have been unified into a single force. First gravity separates out leaving the other three forces as a 'Grand Unified Force'. When the strong nuclear force separates from the 'electroweak' force $10^{-35}$ seconds after the Big Bang, inflation begins. Following the end of inflation, the matter in the universe consists of a soup of quarks and leptons, but the dominant form of energy is radiation. When the universe is a hundred thousand millionth of a second old, the weak nuclear and electromagnetic forces separate and when it is a microsecond old, the quarks bind together to make protons and neutrons. At 1 second, nucleosynthesis begins and continues until the universe is about 3 minutes old. When the universe is 300 000 years old, the universe becomes transparent to radiation and galaxies and clusters of galaxies begin to form.

epoch. So before this epoch we can think of the universe as a fairly smooth gas.

When the universe was 1000 times smaller than at present, the temperature was about 3000 K. An important change to the matter occurs at this point, because the most abundant element, hydrogen, can no longer hold on to its electron at this temperature. The gas becomes

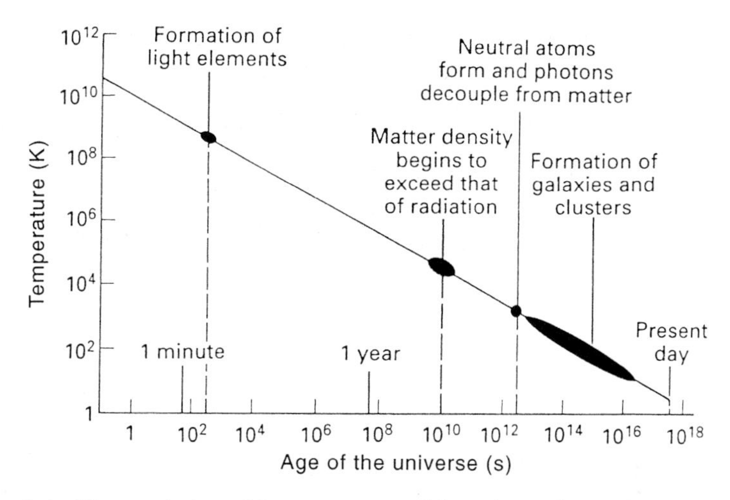


**Fig. 5.4** Time evolution of the temperature of the universe from a time one second after the Big Bang to the present. As the universe expands, its average temperature and density both fall steadily.

'ionized' and consists of electrons moving around freely, and nuclei (mainly protons, with a few helium nuclei). The free electrons have a dramatic effect on the radiation. Whereas after this phase the photons that make up the background radiation travel through the whole universe rarely encountering any matter, before it they are scattered by the electrons very frequently, in a process discovered by the British physicist J. J. Thomson. The universe becomes like an impenetrable fog, through which light can not pass, but can only rattle around to give a uniform haze. Because this phase is the moment when negatively charged electrons become attached to positively charged protons to make neutral hydrogen atoms, it is known as the **recombination era**, by analogy with the process in a cooling ionized gas in our Galaxy. It is not a good name because the electrons were never previously attached to protons so there is no 're' (Latin: back or again) about it. The recombination era occurred about 300 000 years after the beginning of the Big Bang. When we look outwards at the microwave background radiation today we are looking back to this moment in the history of the universe. It is the limit of our observable universe using light and defines a kind of horizon for our observations with telescopes. However, we have other probes which allow us to study still earlier stages of the Big Bang.

## Cosmological nucleosynthesis

We continue our journey back in time and return to the stage where the universe was 400 million times smaller than today, so the temperature was about a billion degrees Kelvin. At this time we will find that temperatures and densities (for all the time that the universe is contracting, the average density is getting steadily higher) are sufficiently high for nuclear reactions to be taking place. Protons and neutrons are fusing together to make the light elements deuterium, helium, and lithium. The whole nuclear-burning phase lasts only a few hundred seconds (a good account of this is given in Steven Weinberg's 1977 book *The first three minutes*). Before 100 seconds after the Big Bang the temperature is too high for deuterium, which is the first step in the nucleosynthesis ladder, to survive, so the reactions do not get started. After a few hundred seconds the temperature has dropped too low for nuclear reactions to continue (nuclear reactions require very high temperatures so that the nuclei are moving around very fast and can collide with each other head on and not simply be repelled by electrostatic forces). The fraction of the neutrons and protons around at 100 seconds which get converted into deuterium, helium, and lithium depends very sensitively on the average density of baryons (neutrons + protons) at this time. We can turn this the other way round. From estimates of the primordial abundances of these elements, that is the abundances before star formation and evolution starts to change the relative abundances of the elements, we can determine the density of baryons that were around when the cosmological nucleosynthesis took place. From this we can estimate the average density of baryons today as $2.5 \times 10^{-28}$ kg m$^{-3}$, with an uncertainty of 20% either way. This was the density we reported in Chapter 1.

Travelling further back in time, at about 4 seconds after the Big Bang another important event occurs. Prior to this moment there were huge numbers of electrons and their antiparticle, positrons (the same mass, but positively charged). At earlier times the photons of the radiation field have so much energy that they can convert spontaneously into an electron–positron pair, via Einstein's famous equation

$$E = mc^2,$$

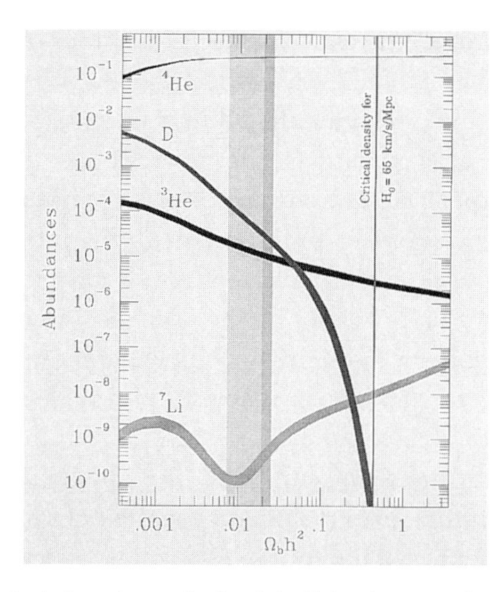


**Fig. 5.5** Cosmological nucleosynthesis of the light elements, showing the abundance of deuterium, helium, and lithium as a function of the density of baryons in the universe. The shaded band shows the range permitted by the observed primordial abundances.

which tells us that energy and matter can be interchanged (when conditions are right). At very early epochs the number of electrons and positrons is about the same as the number of photons and all are in 'thermal equilibrium' together; that is, each electron and positron carries about the same amount of energy. The photons of radiation also each have about this same amount of energy, which is determined by the temperature. Once the radiation cools below 10 billion degrees, the photons no longer have enough energy to make an electron–positron pair and so most of these pairs start to annihilate. If there had been exactly the same number of electrons and positrons, they would all have annihilated and there would be no matter in the universe today, just radiation (protons and neutrons, or more precisely their constituent quarks (see below) would also have annihilated with their antiparticles at an earlier time). For reasons that we do not yet know for certain, the universe happened to have a small excess of matter (electrons) over antimatter (positrons), so when all the positrons had been annihilated there were still some electrons left (and an equal number of protons so that the universe has a net electric charge of zero—otherwise electrostatic forces would be overwhelmingly stronger than gravity). The most likely explanation of the origin of the excess of baryons over anti-baryons (and leptons over anti-leptons) lies in the 'Grand Unified era', when the nuclear forces and the electromagnetic force were all unified into a single force (see below).

The baryon excess gives us another interesting cosmic number: the number of photons per baryon (i.e. protons + neutrons), which stays constant from this moment on. This has a value of about a billion, which shows how close the universe was to complete annihilation of matter (one part in a billion of the electrons were left after the electron–positron holocaust). This is a highly significant number and of great interest to theoretical cosmologists, but in the context of this book it is not a new number, because it can be derived as a combination of the temperature of the microwave background radiation (our fifth number) and the density of baryons (our first number). The number of photons per baryon is in fact proportional to the cube of the microwave background temperature divided by the density of baryons ($T_0^3/\rho_b$). So if I wanted to stick strictly to the policy of using only dimensionless numbers, I would have chosen this rather than the temperature of the microwave background radiation, $T_0$. However, the number of photons per baryon (also referred to as the entropy per baryon) is a rather abstract idea, not as easily grasped as a temperature. Use of the entropy per baryon would also have obscured the incredible accuracy to which we now, thanks to COBE, know $T_0$. The number of photons per baryon is the number which characterizes the degree of asymmetry we have in our universe between matter and antimatter, an issue of great significance to ourselves. Our existence hinges on this process during the Grand Unified era which generated a minute, one part in a billion, excess of matter over antimatter.

## Cosmological neutrinos

As we go back further in time we probe deeper into the mysteries of particle physics. So far we have talked about protons, neutrons, and electrons, the building blocks of the atom. Now other particles come into play. The first is the neutrino, a type of particle which has no charge, and in the basic model for fundamental particles used by particle physicists, known as the **standard model**, is assumed to have no mass (but see Chapter 7). This makes it rather like the photon, the particle of light. However, unlike the photon, the neutrino hardly interacts with matter at all and is therefore very difficult to detect. It was discovered through detailed study of the process of radioactive $\beta$-decay, in which a neutron decays into a proton and an electron. In 1930 Wolfgang Pauli noticed that there is an energy imbalance in this reaction: the neutron always carried slightly more energy than the sum of the energies in the proton and the electron. He suggested that there must be an additional particle carrying this energy away, the neutrino. We now believe there are three types of neutrino: the electron neutrino (involved in $\beta$-decay), the muon neutrino, and the tau neutrino. The latter is not yet detected but is presumed to exist according to the standard model of particle physics, which assumes that each lepton (electron, muon, tau) will have an associated neutrino. Neutrinos are emitted quite profusely by the sun, from the nuclear reactions in its deep interior. When they arrive at the earth, travelling at the speed of light, most pass right through the earth without any interaction. A few can be captured in subtle experiments and there are now several of these in operation around the world, for example the giant Kamiokande experiment in Japan. A blast of neutrinos is also emitted when a supernova explodes and these were seen when supernova 1987A in the Large Magellanic Cloud, our closest known supernova for centuries, was seen to explode in January 1987.

Although neutrinos interact only weakly with matter under terrestrial conditions, in the dense, hot conditions of the very early universe they interact strongly with the matter and radiation and are in equilibrium with the radiation. At 1 second after the Big Bang, when the temperature was 10 billion degrees Kelvin, this interaction ceases and electron neutrinos decouple from everything else. They start to move freely through the universe, like the photons do after the era of recombination. In a further complication the electron neutrino has an antiparticle, the electron anti-neutrino, which differs from the neutrino only in having an opposite **spin**. Apart from photons, all the most fundamental particles have a property which is called spin, in addition to the more obvious ones of mass and charge.

We have by no means reached the limit of the physics we can probe in terrestrial particle physics experiments, though we are moving into the realm of the giant particle accelerators, where particles are accelerated to the enormous energies encountered in nature during the early stages of the Big Bang. At one millionth of a second after the Big Bang, when the temperature was 10 million million degrees, we have the moment when protons and neutrons were assembled. According to the standard model of particle physics, protons and neutrons are not in fact fundamental particles but are each made up of triplets of **quarks**. All together there are six quarks, and these have been given the arbitrary and colourful names 'top', 'bottom', 'up', 'down', 'strange', and 'charmed' quarks. The proton is composed of two up quarks and one down quark, and the neutron of one up and two down quarks. The idea that quarks are the building blocks of matter was first proposed by Murray Gell-Mann and George Zweig of the California Institute of Technology in 1964. Experimental evidence for the existence of quarks began to appear in the late 1960s in experiments at SLAC, the Stanford Linear Accelerator. Today, the masses of all six quarks have been measured.

Prior to one millionth of a second after the Big Bang quarks and their antiparticles exist in huge numbers, in thermal equilibrium with the radiation. At this point most of the quark pairs annihilate and the residue combine together into protons and neutrons. The phase of the Big Bang before this epoch is known as the 'hadron era' (hadrons means heavy particles like protons and neutrons, which are composed of quarks, in contrast to leptons, light particles like electrons and neutrinos). In the hadron era, quarks, electrons, and neutrinos, and their antiparticles, are all in equilibrium with the cosmic radiation.

## The unification of the forces of physics

We have still not quite discussed all that has been learnt from particle accelerators on earth. The most powerful accelerators are able to recreate for an instant conditions as they were in the universe one hundred thousand millionth of a second after the Big Bang. And at that moment an important change of state in the universe occurred. To understand this we have to look at the forces of modern physics. At the start of the nineteenth century three main physical forces were known: electricity, magnetism, and gravity. James Clerk Maxwell showed that electricity and magnetism are two sides of the same coin: the electromagnetic force. What looks like an electric field in one frame of reference will seem to be a magnetic field in another. Then, in the 1890s radioactivity was discovered and this led to two new forces, corresponding to the two types of radioactive transmutation. In $\alpha$-radioactivity, the nucleus of an atom splits and $\alpha$-particles, which are in fact helium nuclei, are ejected. This is a manifestation of the strong nuclear force which holds together the protons and neutrons in an atomic nucleus. $\alpha$-radioactivity is one example of the more general phenomenon of fission, the splitting of the atom, which drives the world's nuclear reactors. The second type of radioactive decay, $\beta$-radioactivity, is a less violent process in which a neutron in an atomic nucleus decays into a proton and an electron. As we saw above, excess energy in this process is carried off by another particle, the electron neutrino. An example of $\beta$-radioactivity is the decay of the radioactive isotope carbon-14 to nitrogen. The carbon-14 nucleus has six protons and eight neutrons. The transformation of one neutron into a proton leaves seven protons and seven neutrons, which is the nucleus of a normal nitrogen atom. The process which controls $\beta$-decay is the weak nuclear force. Both the weak and strong nuclear forces are very short range, operating only within an atomic nucleus.

So today we have four forces of physics: gravity, electromagnetism, the weak nuclear force, and the strong nuclear force. In 1971 Steven Weinberg of the Massachusetts Institute of Technology and Abdus Salam of Imperial College proposed that, under the extreme conditions which would exist in the early universe, the electromagnetic and weak nuclear forces would have been unified into a single 'electroweak' force. This prediction was confirmed in accelerator experiments at CERN in Geneva in 1989 through the detection of the W and Z particles which are involved in the action of this force. In the early universe the transition from the electroweak unified force to two separate forces would have occurred one hundred thousand millionth of a second after the Big Bang. Such a change is called a phase transition, by analogy with the changes that occur in ordinary materials at particular temperatures, for example the change from water to ice at zero degrees Celsius.

Although we have reached the limits of what we know from modern experimental particle physics, theoretical ideas can take us back much closer to the Big Bang. It is natural to consider next whether there was a time when the strong nuclear and electroweak forces were unified into a single force. Extrapolation of what we know about these forces suggests that this must have happened prior to about

$10^{-35}$ seconds after the Big Bang. This is a number so small, 0.00000 … with 35 zeros … 001, that there is no point in trying to give it a name in terms of millionths and billionths. At this time what has become known as the Grand Unified Force split into the electroweak and strong nuclear forces. Even though we do not have a completely satisfactory theory for this transition and the most obvious prediction of this type of theory, that the proton would decay over a long timescale, has not yet been confirmed, there is still a widespread belief among physicists that the idea is correct. As we have mentioned above, this is also believed to be the moment when the symmetry between baryons and their antiparticles was broken, generating the small excess of baryons over anti-baryons which ultimately resulted in the baryon-dominated universe we see today.

The final process of unification, of gravity with the Grand Unified Force, would have taken place at around the Planck time, $10^{-43}$ seconds after the Big Bang (see p. 63). Currently there is a lot of interest among theorists in trying to characterize what a quantum theory of gravity should be like, but these 'Theories of Everything' remain extremely speculative.

## Phase transitions, defects, and inflation

So we expect that there will have been at least three major phase transitions in the early universe, at the Planck time, at the Grand Unified epoch, and at the electroweak epoch. There may also have been others, associated with particles and forces that we do not yet know about. These transitions offer interesting possibilities for shaking the universe up a bit and leaving traces of their passing. In ordinary terrestrial materials a consequence of phase transitions is that they can leave defects or fault lines in the material in which they occur. This same idea was applied to phase transitions in the universe by Tom Kibble of Imperial College in 1976. He predicted that defects generated by phase transitions in the early universe could play an important role in cosmology. He classified defects into three main types: 'monopoles' (a point-like defect), 'cosmic strings' (line-like), and 'domain walls' (plane-like), each of which would be localized regions of very high energy density. He suggested that in particular cosmic strings could play a part in generating the structure we see in the universe today. This is a very interesting idea which is still being actively pursued by several groups around the world.

A more drastic consequence of a phase transition is that the universe could have gone through a period of rapid and exponential 'inflation' during the transition which would have completely transformed its character. The inflation would be driven by an immense energy density associated with the vacuum, which acts like a temporary, very strong cosmological repulsive force. This idea was first proposed by Alan Guth in 1981 to solve what he called the 'horizon problem', and some other difficulties with the standard Big Bang model. Most but not all inflationary universe models associate inflation with the Grand Unified phase transition $10^{-35}$ seconds after the Big Bang. We have mentioned the concept of horizon before. The horizon problem is that when we look at the microwave background radiation in two opposite directions in the sky, the emitting regions are separated from each other by 50–100 times the size of the horizon at that time, so in the standard model they would never have been able to communicate with or influence each other. We say that they were not in 'causal' contact with each other, since one region can not cause anything to happen in the other. How then did they come to be so similar to each other? The standard model has to push the processes that generated the isotropy of the universe back to the Planck time, essentially to initial conditions. In the inflationary model, the two regions would have been in causal contact prior to the inflation epoch so there is a causal explanation of the isotropy of the universe.

Other 'problems' solved by inflation are not so compelling. The 'flatness' problem is based on the fact that if the average density of the universe were enormously different from what is observed, then we would not be here. If the universe were too dense the expansion would already have halted long ago and the Big Crunch already have occurred. If the density were too low galaxies could not have formed. So it looks as if some kind of 'fine-tuning' is required to get the universe we observe. This, however, does not seem genuinely paradoxical in the way that the horizon problem does. It seems obvious that the universe we are in has to be such that galaxies, stars, and we ourselves can form. Alan Guth was also interested in the capacity of inflation to get rid of excessive numbers of monopoles likely to be generated in the Grand Unified phase transition through the mechanism proposed by Kibble.

The actual inflationary mechanism proposed by Alan Guth did not work, and a more sophisticated version was produced by Andre Linde, and by Andy Albrecht and Paul Steinhardt in 1982. A very important by-product of inflation is that as the inflationary period comes to an end and the energy of the vacuum is converted into matter and radiation, small density fluctuations would be generated of about the right form and strength to make galaxies and clusters in the later stages of the universe. Without inflation, these fluctuations would have to be made back at the Planck era.

We will return to many of these questions. We see that from the simple discovery of microwave background radiation we have constructed a weird and wonderful story for the early universe which takes us far beyond the physics we really know. It is fortunate that this edifice is built on the foundation of the best-determined of all of our cosmic numbers: the temperature of the background radiation, $T_0$.

Chapter 6

# Cold dark matter

*These people, however, are unaware of anything other than perceptible matter.*

Aristotle, *On the heavens*

The processes by which stars and planetary systems form from interstellar gas clouds, and by which galaxies form from protogalactic gas clouds, are among the most intractable in modern physics. We know a great deal about general relativity, black holes, particle physics, the evolution of the universe. But the details of how stars and galaxies form remain, literally, hidden from view. We shall see that one of the key ingredients to explain the formation of galaxies is a new form of matter, **cold dark matter**. Our sixth cosmic number is the average density of cold dark matter in the universe.

The fundamental idea that underpins all attempts to explain the formation of stars and galaxies is that gravity is the main cause. In 1796, just over a century after Newton's discovery of universal gravitation, Laplace put forward his nebular hypothesis, that the sun and planets formed by gravitational condensation out of a rotating, disc-shaped cloud of gas. The cloud would initially be of uniform density, prevented from falling together by centrifugal force, but instabilities would arise in the cloud which would give rise to the sun and planets. Immanuel Kant had suggested in 1755 that a model of this type could explain the formation of our Milky Way galaxy. Nineteenth-century studies of the stability of rotating disc-shaped gas clouds did not, however, give much insight into how Laplace's nebular hypothesis would work.

## Gravitational instability

In 1902 James Jeans analysed a problem which seemed much more relevant to the formation of galaxies in an initially smooth universe. He studied the stability of an infinite, uniform distribution of gas and found a criterion for a density fluctuation, a region in which the density was slightly higher than average, to keep on getting denser and hence eventually collapse together to form a self-gravitating object like a star or galaxy. The Jeans criterion is quite simple. The time for sound waves to cross the fluctuation must be greater than the time it would take for the region to collapse together under gravity if there were no pressure forces opposing the collapse. The significance of the speed of sound waves for the problem is that this is the speed at which pressure disturbances move in the gas and these are what oppose the tendency of gravity to cause collapse. The time for a mass of gas to collapse under the action of gravity, the 'free-fall time', depends only on the density of the gas. For the sun, the free-fall time is about 1 hour and this is the time it would take the sun to collapse to a black hole if the pressure gradient which is caused by the huge temperature difference between the centre and surface of the sun were removed suddenly. Jeans' calculation gives the minimum size of a cloud which can collapse under the influence of gravity, the 'Jeans length'. For a gas cloud with the average density of the universe, the free-fall time is about the age of the universe, so only a region of size comparable to the whole observable universe is unstable to gravitational collapse. New galaxies are not being formed today.

To find the epoch when galaxies started to form we have to go back to a much earlier stage in the evolution of the universe, to when the observable universe was much smaller than today and contained the mass of about one galaxy. For a galaxy like ours this was about a hundred years after the Big Bang. Now this was a phase when the dominant form of energy was not matter but radiation and this modifies the Jeans argument. We have to do the calculation not in Newtonian gravity, as Jeans did, but in Einstein's general theory of relativity. Finally, we have to allow for the fact that the initial state is not static, as assumed by Jeans, but corresponds to an expanding universe. When all this is allowed for we still, perhaps surprisingly, find a condition very similar to Jeans'.

The rate of growth of perturbations satisfying the Jeans criterion turns out to be much slower in an expanding general relativistic universe than calculated by Jeans. This is exacerbated by the fact that during the early radiation-dominated phase of the universe the Jeans length is only slightly smaller than the causal horizon of the universe and increases with time at the same rate as the horizon. So a perturbation finding itself inside the horizon for the first time starts to contract and increase its density. But after a short time the Jeans length has increased to be larger than the perturbation, so the condition for growth is no longer satisfied and the perturbation remains frozen until the era of recombination. At this point the Jeans length shrinks enormously and the perturbation is free to start to contract under gravity again and become a galaxy.

## How strong were the density fluctuations?

At this point, to decide whether a density fluctuation containing enough mass to make a galaxy will in fact do so by today, we only need to know by how much its density differs from the average density of the universe. A fluctuation with a strong density excess will form quickly; one whose density is only very slightly higher than the average may not form in time. Bear in mind that the gravity of the fluctuation has to overcome the expansion of the universe because initially everything is moving with the expanding universe. It turns out that to form by today, a density fluctuation must exceed the average density by at least three parts in a thousand (0.3%) at the era of recombination. This may not seem very much but it is rather a large deviation from smoothness if the universe started off completely smooth.

We saw in the previous chapter that when Penzias and Wilson discovered the microwave background radiation, they found that it had the same brightness in whichever direction in the sky they looked. By the late 1970s the radiation was known to be isotropic, the same in every direction, to one part in ten thousand, apart from the dipole anisotropy due to our motion through space. Since matter and radiation were locked together up to the era of recombination, and the density is changing as the cube of the temperature, the density fluctuations at this epoch could be at most three parts in ten thousand. These limits on the temperature fluctuations in the microwave background radiation seemed to rule out the idea that galaxies were made by gravity acting on density fluctuations arising in the early universe. Cosmologists tried to think of ways to get round this or of other ways that galaxies might be formed. One idea was that the radiation was completely smooth, as observed, but that the matter was lumpy—these are called 'isothermal' fluctuations. This idea was considered too artificial to be a runner. Even if the universe started off like this, fluctuations in the radiation would soon be generated. A second way out was to suppose that the required fluctuations were really present but masked because an excess in matter density was compensated for by a corresponding deficit in radiation density, called 'isocurvature' fluctuations. These are a bit harder to rule out but are not considered likely by most cosmologists. Another idea for making galaxies which evaded the Jeans condition was that some kind of cosmic explosions scattered through the universe would compress gas together and trigger its collapse. The problem then is shifted back to one of making whatever objects it is that explode.

## Hot and cold dark matter

Cosmologists began to consider more radical alternatives. Perhaps density fluctuations of the required strength really were present at the era of recombination but were fluctuations in some kind of matter which no longer interacted with the radiation. The idea of dark matter was born. One example of a particle which ceased to be in thermal equilibrium with the radiation long before the era of recombination is the neutrino, which we encountered in the previous chapter. Normally a neutrino would be no help in the process of galaxy formation because it has no mass so can exert no gravitational attraction. But some particle physicists had been playing with the idea that the neutrino might have a small non-zero mass (this mass may now have been detected, as we shall see in Chapter 7). If the mass of the neutrino was about one twenty thousandth of the mass of an electron then neutrinos would be the dominant form of mass in the universe. If the Jeans calculation is redone for a mixture of radiation, ordinary ('baryonic') matter, and neutrinos, the outcome is quite different. In a universe dominated by neutrinos only very large fluctuations with masses of thousands of galaxy masses can start to contract. This is because the neutrinos move around at speeds close to that of light and quickly escape from smaller density fluctuations. This restriction to very large density fluctuations suggested that the sequence of galaxy formation would be that giant clusters of galaxies formed first and that when the cloud of gas collapsed together to make the first large structure, galaxies would then form. The Russian cosmologist Yakov Zeldovich explored these ideas in some detail and suggested that these large structures would in fact collapse to form a flat 'pancake'. The gas would be heated up strongly, would radiate and cool down, and then instabilities in the gas would form the galaxies.

To see whether this idea would really work, in 1983 Simon White, Carlos Frenk, and Marc Davis, all then working at the University of California at Berkeley, set out to make computer simulations of this process. They found a severe drawback to the idea of a neutrino-dominated universe. Because the largest structures formed first, these models tended to give a universe with too much structure on large scales. Although we do see rich clusters of galaxies out in the universe, only about 10% of the galaxies are in such structures. Most are in rather weaker groupings and clusterings. The neutrino-dominated universe did not seem to give this kind of picture.

If the dark matter particle were a neutrino, then in the early universe it would be moving around at a speed very close to the speed of light. Such dark matter particles are called 'hot' dark matter. It is these high velocities which make it hard for neutrinos to cluster together on any but the very largest scales. An alternative is to postulate particles that move around very slowly in the early universe. These would be 'cold' dark matter. At the time these ideas were being explored, in the late 1970s, there was no obvious candidate for a cold dark matter particle. But theorists pressed ahead with exploring what a universe with a large population of cold dark matter particles would be like. Cold dark matter particles would tend to cluster together on scales smaller than galaxies. Globules of cold dark matter would condense and start to merge together under the action of gravity, gradually building up to galaxy-sized masses. Meanwhile, the much more smoothly distributed ordinary baryonic matter would, once the era of recombination was past, begin to feel the gravitational pull of the nearest concentrations of cold dark matter and start to fall towards them. Simulations suggested that the cold dark matter would end up as the haloes of galaxies with the baryons concentrating towards the

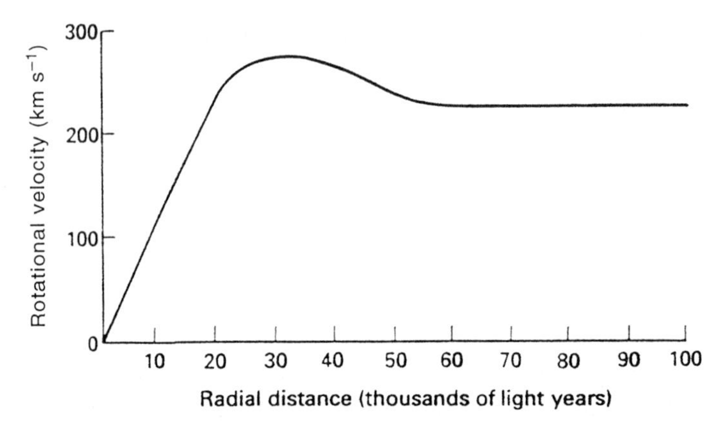


**Fig. 6.1** Rotation curve of the spiral galaxy M31. The rotational velocity does not fall off with radius in the outer regions as would be expected if there were only the visible matter in the galaxy disc. The almost constant value of the rotational velocity implies there is a halo of dark matter surrounding the disc.

centres of these dark matter haloes to form the visible distributions of stars and gas that we see in galaxies.

A great attraction of this idea was that our Galaxy and several others were being found by astronomers to be surrounded by massive dark haloes of unknown origin (see next section). The cold dark matter universe seemed to solve this problem. Moreover, when computer simulations of these models were made, for example by Marc Davis and his collaborators, they seemed to correspond rather well with the observed galaxy distribution.

## The spectrum of density fluctuations on different scales

A key assumption which has to be fed into these computer simulations of galaxy and cluster formation is how strong the density fluctuations are on different scales at the start of the calculation. What we really want to know is how the average value of the fractional density fluctuation ($\Delta\rho/\rho$) depends on the length scale (or alternatively on the mass in the fluctuation). This function is called the density fluctuation spectrum. We could start the calculation, for example, from the era of recombination. The COBE measurements give us a good idea about the strength of the density fluctuations on the very largest scales, characterized by the second of our cosmic numbers, $\Delta T/T$. The COBE measurements also suggest that $\Delta\rho/\rho$ increases as we look towards smaller scales. More precisely, we find that $\Delta\rho/\rho$ increases approximately as the inverse square of the length scale. Now this form of spectrum for the density fluctuations has a very natural interpretation. Remember that the radius of the horizon of the universe is steadily increasing with time, so that whatever length scale you choose (smaller than the radius of the horizon today) there was a moment in the past when the radius of the horizon was equal to that length scale. It turns out that at the time when any particular fluctuation first came inside the horizon of the universe (so the length scale of the perturbation is equal to the size of the horizon), then the value of $\Delta\rho/\rho$ at that time would be the same whatever scale we consider. A spectrum of this kind had been proposed in 1970 by Edward Harrison of the University of Massachusetts and in 1972 by Yakov Zeldovich of the Academy of Sciences, Moscow, as the most natural for the universe, because it involves no characteristic length scale (no minimum or maximum or typical galaxy mass). When we look at the clustering of galaxies today we do seem to see clustering on all scales, with no typical size for clusters, as proposed by Harrison and Zeldovich. It seems natural to assume that the type of spectrum of density fluctuations proposed by Harrison and Zeldovich, and eventually seen by COBE, extrapolates down to all length scales and this is what was assumed in the computer simulations. The success of the cold dark matter models in generating, from very simple assumptions about the initial distribution of density fluctuations, a distribution of galaxies and clusters very similar to what we observe, resulted in strong support for these models which has persisted to the present day. There have been a few hiccups along the way which meant that the theory has had to be adjusted, and in the next chapter we will look at some of these refinements more closely. However, the view prevails, almost as strongly as the idea that galaxies are formed by the action of gravity on primordial density fluctuations, that the universe is pervaded by cold dark matter and that it makes up the bulk of the dark matter in the halo of our Galaxy. Thus, the sixth cosmic number is the average density of the universe in the form of cold dark matter, which we measure in dimensionless form by $\Omega_{cdm}$.

In the simplest picture the total density of matter in the universe would then be the sum of the density in ordinary baryonic matter and the density in cold dark matter. Since we know the density parameter for baryonic matter quite accurately, $\Omega_b = 0.03 \pm 0.006$ (Chapter 3), if we could measure the total density of matter in the universe, we could then deduce the density of cold dark matter by taking the difference. The simplest model of the universe is the model of Einstein and de Sitter, which is spatially flat and in which the density has the critical value (see p. 57) and the universe keeps on expanding for ever, but with the rate of expansion getting ever slower as time increases. In this model the value of the total density parameter $\Omega_{tot}$ would be 1 at all times. Such a value is also supported by advocates of inflationary models since inflation is expected to drive the curvature of the universe so close to zero that it would still remain spatially flat at the present day. If this model is correct then for a universe with baryons and cold dark matter only, we would have $\Omega_{cdm} = 0.97$, so 97% of the universe would be in the form of cold dark matter and the ratio of cold dark matter to ordinary matter would 32:1. We will return to the question of what is the total density of matter in the universe in the next chapter. There is still controversy over the value of $\Omega_{tot}$, with estimates ranging from 0.1 to 1 (see Chapter 7).

## Dark haloes: machos or cold dark matter?

When we look at our own Galaxy and other nearby galaxies which have been studied in detail, we find that the dark haloes of the galaxies account for about ten times as much matter as we can see in the stars and gas of the galaxies. The mass of the dark halo of our own and other spiral galaxies is estimated by studying how fast the stars are orbiting round the galaxy, which can be measured using the Doppler shift. If the mass distribution in a galaxy followed that of the visible stars, we would expect the orbital speed to be slower in the outer parts of the Galaxy, as it is for the outer planets of the solar system. In fact once we are so far out that the stars are no longer adding much to the galaxy's total mass, we would expect Kepler's law to hold: that the orbital speed falls off as the inverse square root of the distance. This can be derived by just balancing the centrifugal force acting on the orbiting star against the gravitational pull of the galaxy. In fact what we see in galaxies is that the orbital speed of stars remains almost constant as distance increases, which implies that there must be a lot more mass in the outer parts of galaxies than can be seen in visible stars. The dark haloes of our own and other galaxies have been known for 30 years

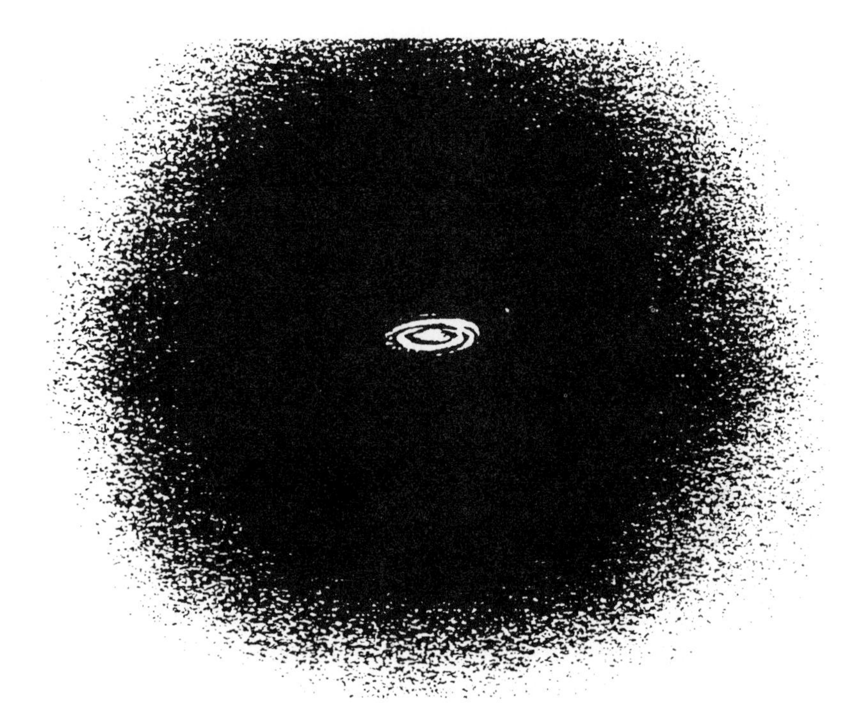

**Fig. 6.2** Artist's impression of a dark halo surrounding a spiral galaxy.

and were first discovered in 1970 by Mort Roberts of the US National Radio Astronomy Observatory, Vera Rubin of the University of Washington, Ken Freeman of Mount Stromlo Observatory, and others, through studies of the rotation curves of galaxies at radio and optical wavelengths.

What could this dark matter be? Recent experiments looking for gravitational lensing effects due to stars in the halo of our Galaxy suggest that at least some of the dark halo is due to stars. These experiments rely on monitoring closely the brightness of millions of stars in a nearby external galaxy like the Large Magellanic Cloud (see p. 26). If a dark star in the halo of our galaxy passes across the line of sight to one of these background stars, the gravitational bending of light causes an apparent brightening of the light from the background star, the phenomenon of 'microlensing'. Several dozen good examples have been seen, though it is not clear yet what the lensing stars are. They seem to have a mass of between a tenth and a half of the mass of the sun. If these lensing objects are in the halo of our Galaxy they can not be normal hydrogen-burning stars or we would be able to see them in deep optical surveys, so the best candidates are white dwarf stars, dead remnants of stars like the sun and much less luminous. However, no one has come up with a convincing story of how there could be so many dead white dwarf stars in the halo of our Galaxy. Estimates of the proportion of the dark halo contributed by the microlensing stars range from 10 to 50%, with a best bet of about 20%. If we suppose the rest of the dark halo is due to cold dark matter, then the ratio of cold dark matter to ordinary baryonic matter in our Galaxy is about 3:1, a lot less than the 32:1 mentioned above. This might not be surprising because there could be plenty of cold dark matter which was not associated with galaxies. Another possibility which looks increasingly interesting is that the lensing stars may also be in the Large Magellanic Cloud; that is, one star in the LMC is lensed by another star in the LMC passing in front of it. Two cases of binary lensing systems have been found and for these we can deduce a distance of the lens. In both cases they turn out to be at the distance of the Large Magellanic Cloud. If this turns out to be the right interpretation then the ratio of cold dark matter to baryons in our Galaxy would increase to 10:1.

However, another estimate of this ratio can be made in rich clusters of galaxies. We first add up the baryonic mass of all the galaxies in the cluster by measuring the total amount of starlight in the galaxies and then using the typical ratio of starlight to mass found in our own Galaxy and other nearby galaxies. We then add to this the mass of the hot X-ray-emitting gas in the cluster. We saw in Chapter 3 that rich clusters of gas are permeated with very hot (100 million degrees Celsius) gas which radiates at X-ray wavelengths. Finally, we estimate the total mass of the cluster, including its dark matter, by calculating how much mass is needed for the galaxies to be moving around at the speeds observed. The total mass of the cluster can also be estimated independently by studying the lensing arcs which are seen in many clusters. We find that about one third of the total mass of the cluster is contributed by ordinary baryonic matter. So both in our Galaxy (if the lensing objects are white dwarfs) and on the much larger scale of rich clusters of galaxies, about 25–35% of the total mass is due to baryonic matter, with the remainder presumably due to some kind of non-baryonic dark matter, for example cold dark matter. These arguments suggest we should multiply the average density of baryonic matter in

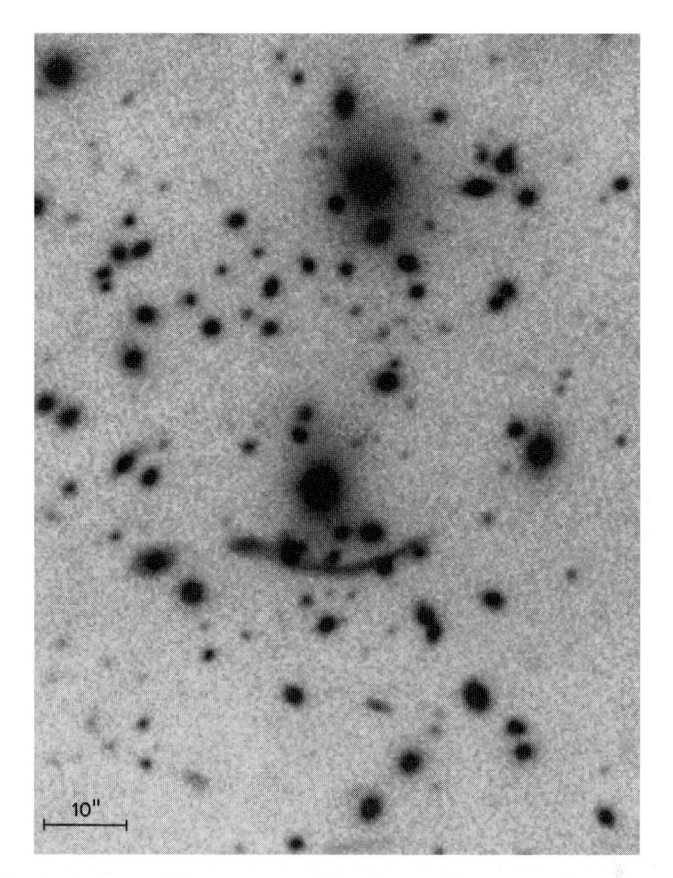


**Fig. 6.3** A gravitational lens arc in a galaxy cluster. These arcs can be used to estimate the mass of dark matter in the cluster.

the universe by about 2–3 to get the density of cold dark matter, which would give $\Omega_{cdm} \approx 0.06$–$0.09$. If the microlensing objects are normal stars in the LMC, this estimate increases to $\Omega_{cdm} \approx 0.3$.

## The nature of the cold dark matter

What could the cold dark matter be? The current most popular idea is that it is a particle postulated by particle physicists in their efforts to unify the forces of physics. In order to make different types of fundamental particle like quarks and electrons equivalent to each other, theorists postulate the concept of 'supersymmetry', which requires there to be many new particles, essentially one new particle for each of the known particles (denoted by adding the letter s before, or the suffix -ino after, the name of the particle—squarks, selectrons, sneutrinos, photinos, etc.). Most would be far too massive to occur naturally or even under accelerator conditions. However, the lightest of these supersymmetric particles, the 'neutralino', might have a mass around 100 to 1000 times the mass of a proton and could exist in sufficient numbers to be the cold dark matter particle. There are other particle physics candidates for cold dark matter, for example the 'axion', a particle postulated in certain models to fix up problems with the standard theory of strong interactions, which is called quantum chromodynamics or QCD. A completely different possibility for cold dark matter is that it consists of 'primordial' black holes of, say, 1 to 1000 times the mass of the sun, which are supposed to have existed in the universe since very early times. Such an idea is hard to rule out but there does not seem to be a very convincing explanation of how they have been formed.

At least a dozen groups around the world are trying to detect the cold dark matter particles in the halo of our Galaxy, assuming that they are weakly interacting massive particles (WIMPs) like the neutralino. If so, these particles would run into the earth all the time, generally just passing through anything they encountered without interacting. However, they do have a very small probability of hitting the nucleus of an atom and some ingenious methods of detecting the recoils of nuclei as they are hit have been devised. These experiments have to be located deep underground to get away from the pervasive effect of cosmic rays, electrons, and nuclei moving close to the speed of light which bombard the earth all the time and which are generated in the sun and probably in pulsars and supernova remnants.

Earlier experiments were based on attempts to detect electric charges in germanium crystals due to liberation of electrons by the passing WIMPs. Experiments of this type were run by a USA–Argentina collaboration operating at Sierra Grande, Argentina, and by a Heidelberg–Moscow collaboration in the Gran Sasso laboratory under the Italian Alps. The UK has a Dark Matter Experiment running in the Boulby potash mine in Yorkshire, which is over 1 kilometre deep. This illustrates the way that astronomers try to occupy all niches to observe the universe, from mountain tops to telescopes in orbit to the deepest mines. The UK experiment is designed and managed by groups at the Rutherford Appleton Laboratory, Imperial College, and Sheffield University. The mine owners provide two caverns for the experiment and supply electricity, ventilation, and transport. The current detectors consist of large crystals of sodium iodide. When a nucleus is hit by a heavy particle, the atoms emit a flash of ultraviolet light, the phenomenon of fluorescence. Very sensitive light detectors are used to observe these flashes. So far the events seen are all consistent with being due to radioactivity from the walls of the cavern and in the equipment. The limits that can be set on the rate of dark matter collisions are beginning to be interesting and to be at the upper end of the range predicted by theoreticians. So a few more extreme predictions are beginning to be ruled out and if the sensitivity of the experiments can be improved by a factor of ten or so, there is a good chance that the neutralino will, if it exists, be detected.

Although the UK experiment has for several years been the most sensitive cold dark matter search in the world, other experiments are now catching up. An Italian group (DAMA) led by the University of Rome is running an experiment with much larger crystals of sodium iodide and is already achieving even better sensitivities. Both groups have found unexplained anomalies in their data, which may turn out to be due to local radioactive contamination. Detection of the

**Fig. 6.4** The UK Dark Matter Experiment at the Boulby potash mine in Yorkshire.

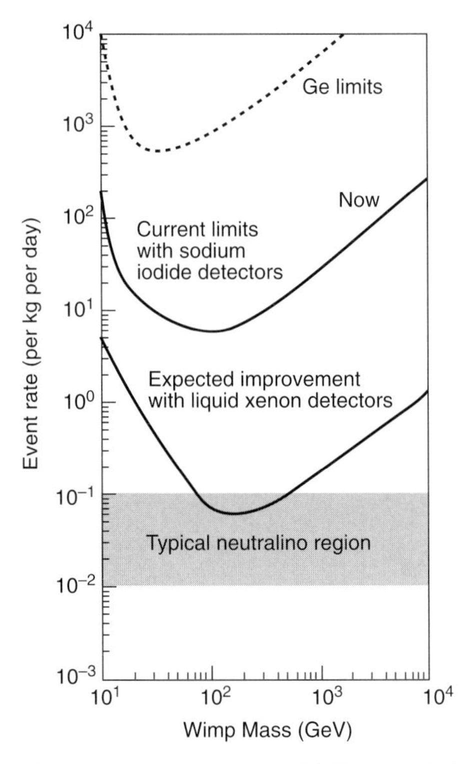


**Fig. 6.5** Current limits on event rates due to WIMPs in the halo of our Galaxy, and the expected improvement by 2003.

neutralino is not really expected until sensitivities are improved substantially. An experiment planned by the University of California at Berkeley and the University of Stanford will use a completely different detection technique based on observing temperature changes in cooled germanium crystals. The UK team are developing a new detector based on liquid xenon, a rare element which exists in gaseous form in the earth's atmosphere, and they hope to improve their sensitivity by a factor of a hundred over the next few years. If cold dark matter dominates the halo of our Galaxy, and if the cold dark matter particle is the neutralino, there is a good prospect of detecting this particle over the next decade. The Large Hadron Collider currently being built at CERN in Geneva is also capable of proving the existence of the neutralino, though unlike the underground experiments it will not be able to show that these particles are present in our Galaxy, since any particles it detects will have been manufactured in very energetic collisions in the accelerator.

The detection of cold dark matter will have a tremendous impact on both cosmology and particle physics. It seems extraordinary that we might be on the brink of finding this invisible but dominating form of the matter in the universe so soon after it was proposed as the solution to the galaxy formation and galactic dark halo problems.

Chapter 7

# The missing ingredient—tilt, strings, or hot dark matter

*Consider, therefore, this further evidence of bodies whose existence you must acknowledge though they cannot be seen.*

Lucretius, *On the nature of the universe*

The invention of cold dark matter at the end of the 1970s seemed to solve many of the problems associated with how galaxies formed. Simulations of galaxy formation in a cold dark matter universe did generate model universes very similar to the observed distribution of galaxies and clusters of galaxies. But when we come to look more closely it seems there is still a problem. It looks as if there is an additional ingredient needed to make sense of the observed clustering of galaxies and in this chapter we consider several possibilities. We still do not know what that missing ingredient might be. The initial spectrum of density fluctuations might be slightly different ('tilted') from the simple form we assumed in the last chapter. The universe might be pervaded by relicts from the inflationary era, of which the most likely are cosmic strings, which we encountered in Chapter 5. Or there might be an additional ingredient of dark matter in the form of hot dark matter, presumably a neutrino with a non-zero mass. In this chapter we discuss these possibilities in detail and explain how they might be detected and quantified. So the seventh cosmic number characterizes the missing ingredient required to understand large-scale structure, which could be tilt, strings, or hot dark matter. A further possibility, to be discussed in Chapter 8, is that there is a non-zero

**cosmological constant**, essentially an additional repulsive force acting on very large scales which can have a big effect on the age of the universe and the evolution of structure.

William Herschel was the first to notice that galaxies are sometimes found in rich clusters when he surveyed the constellation of Virgo in 1784 with his 18-inch reflecting telescope and discovered the Virgo cluster. In *The construction of the heavens* he goes on to describe the most prominent clusters of the northern sky, speculating that they might form a single structure:

> *Another stratum (of nebulae) ... is that of Coma Berenices, as I shall call it ... It has many capital nebulae very near it; and in all probability this stratum runs on a very considerable way. It may, perhaps, even make the circuit of the heavens ... the direction of it towards the north lies probably with some windings through the Great Bear onwards to Cassiopeia; thence through the girdle of Andromeda and the northern Fish, proceeding towards Cetus; while towards the south it passes through the Virgin, probably on to the tail of Hydra and the head of Centaurus.*

Herschel is describing here what we would call today the Local Supercluster, a disc-shaped structure dominated by the Virgo cluster and including the Ursa Major (Great Bear) cluster. In linking this with the clusters in Hydra and Centaurus, which are three times more distant than Virgo, Herschel is foreshadowing the larger local structure which is responsible for much of the motion of our Galaxy through the cosmic frame and which some astronomers like to call the 'Great Attractor'. Herschel is also drawing attention to even more distant clusters of galaxies in the constellations of Coma, Pisces (the Fish), and Cetus, which we would not think of as part of a single structure with the Virgo, Hydra, and Centaurus clusters, unless we were thinking on a very large scale indeed.

However, before we give Herschel too much credit for his large-scale vision, we need to remember that throughout his life he vacillated in his beliefs about whether the nebulae were distant external star systems or much nearer gas clouds in the Milky Way. He was therefore uncertain of the significance of the Virgo cluster and other clusters and larger structures. It was not really till the twentieth century and the pioneering work of the Swiss–American astronomer Fritz Zwicky that the importance of galaxy clusters became clear. Zwicky used the Palomar 24-inch Schmidt telescope to survey the northern sky for clusters of galaxies, as well as for moderately bright galaxies. His six-volume catalogue published in the 1960s, and its extension to the southern sky by P. Nilson in 1973, remains an important tool for research today.

## Large-scale structure and galaxy redshift surveys

The challenge of surveying the large-scale structure of the universe was taken up by astronomers at Cambridge, Massachusetts, first by John Huchra and then by Margaret Geller. In a series of surveys starting in the 1970s with a variety of collaborators they measured the redshifts of thousands of galaxies to get a three-dimensional picture of the universe. They demonstrated the existence of huge sheets of clusters bridging more than one cluster, for example the so-called 'Great Wall' which links the Coma and Hercules clusters. The discovery, by Robert Kirschner of Harvard and colleagues, of a huge void in the constellation of Bootes containing very few galaxies also made a great impact on the astronomical and public consciousness. Were these an intrinsic feature of the galaxy distribution, perhaps with some kind of characteristic length scale associated with them, or were they merely a manifestation of a universal clustering process? The supporters of the cold dark matter picture, for example the theorists Marc Davis of Berkeley, Carlos Frenk of Durham, and Simon White and George Efstathiou at Cambridge, argued that there was nothing special about the sheets and voids. They could be seen in the numerical simulations based on models in which purely random gravitational processes were at work. For the theorists the only important quantity was the spectrum of the initial density fluctuations and this was best measured by studying the clustering of galaxies on large scale. Both observations and theoretical simplicity favoured the assumption that the spectrum of density fluctuations had a very simple dependence on scale, with the density decreasing as a simple power of increasing scale. However, they did have to make one important adjustment to the model. Their simulations predicted that galaxies would be moving around at random much faster than was observed. To fix this they had to assume that the galaxies tended to trace regions of the universe where the density fluctuations were stronger than average, so that the galaxy distribution gave a biased picture of the underlying matter distribution.

## Problems for the cold dark matter model: IRAS and COBE

An important development for the study of large-scale structure was the launch of the Infrared Astronomical Satellite (IRAS) in 1983. I have told the story of this mission in my book *Ripples in the cosmos*. The main goal of IRAS was to survey the whole sky at wavelengths of 12, 25, 60, and 100 microns, the first time the sky had been systematically examined at mid and far infrared wavelengths. The 60-micron survey turned out to be especially important for cosmology because away from the plane of the Milky Way we found that, after exclusion of some obvious stars, virtually all the sources were galaxies. IRAS allowed us to make an all-sky galaxy catalogue that was deeper than Zwicky's optical catalogue, covered both hemispheres, and was free of many of the systematic problems of Zwicky's survey, for example the dimming effect of interstellar dust at optical wavelengths.

I had been closely involved in the preparation of the IRAS catalogues from the satellite data and was confident of their quality. I formed a collaboration with several leading young British astronomers (Andy Lawrence, George Efstathiou, Carlos Frenk, Richard Ellis, and Nick Kaiser) and we set out to measure the redshifts of a large sample of IRAS galaxies. We were fortunate in being given a large block of time—almost four weeks—on the 4-metre William Herschel Telescope on La Palma in the Canary Islands while it was being commissioned in December 1987 and we were able to complete two thirds of the sky during that time. Of course, part of the southern sky is not visible from the Canaries, so we completed the southern hemisphere with observing runs at the Anglo-Australian Telescope at Siding Springs, Australia, the following summer. These redshifts gave us the first ever three-dimensional picture of the galaxy distribution out to significant depths in the universe and this allowed us to do several interesting things. Firstly, we could test whether the net attraction of the galaxy concentrations around us could explain the motion of our Galaxy inferred from the microwave background dipole anisotropy. To our delight the direction in which the galaxy distribution sampled by IRAS was pulling us agreed well with the direction of our motion. To explain the speed of our motion the universe had to have a high total average density, consistent with the critical density (see p. 57), $\Omega_{tot} = 1$. Secondly, we could give the first complete three-dimensional picture of the local universe, including all the clusters and voids. While the clusters were mostly pretty well known, several of the voids were completely new. And thirdly, we could test whether the spectrum of density fluctuations today agreed with the predictions of galaxy formation scenarios like the cold dark matter model. To our great surprise there was a serious conflict with the latter model, which had been so successful in so many other tests. It seemed clear that some other ingredient was needed in the model.

The situation became even more acute when the COBE team announced their discovery of the ripples, which corresponded to density fluctuations on very large scales. The standard cold dark matter model could not fit both the COBE ripples and the galaxy density fluctuation spectrum. The immediate solutions to this problem fell into two camps. George Efstathiou, now at the University of Cambridge, and colleagues advocated introducing a non-zero cosmological constant into the cosmological models. This has the effect of increasing

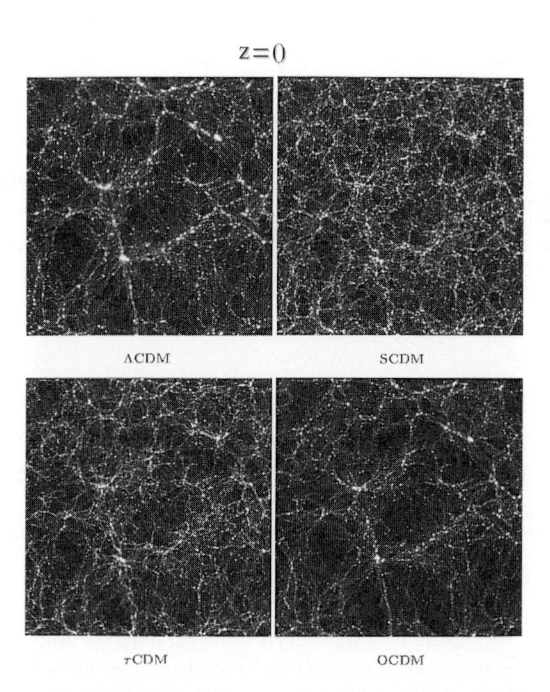


**Fig. 7.1** Computer simulations of the universe today with four different assumptions about the nature of the dark matter. ΛCDM: cold dark matter (CDM) plus a non-zero cosmological constant (see Chapter 8), SCDM: standard CDM, τCDM: CDM with a decaying massive neutrino, OCDM: CDM in the open ($\Omega < 1$) universe (Courtesy of the Virgo consortium.)

the time for fluctuations to grow and this modifies the shape of the fluctuation spectrum as required. We will treat the cosmological constant as a separate issue in the next chapter. My student Andy Taylor and I, and several other groups around the world, preferred models which included a second type of dark matter, hot dark matter. If about 20–25% of the dark matter were in fact due to a neutrino with a non-zero mass, then the shape of the density fluctuation spectrum is changed towards giving more power on large scales as required. This model is called the mixed dark matter model, because it requires both hot and cold dark matter. Simulations of this model were carried out by several groups, with Joel Primack of the University of California at

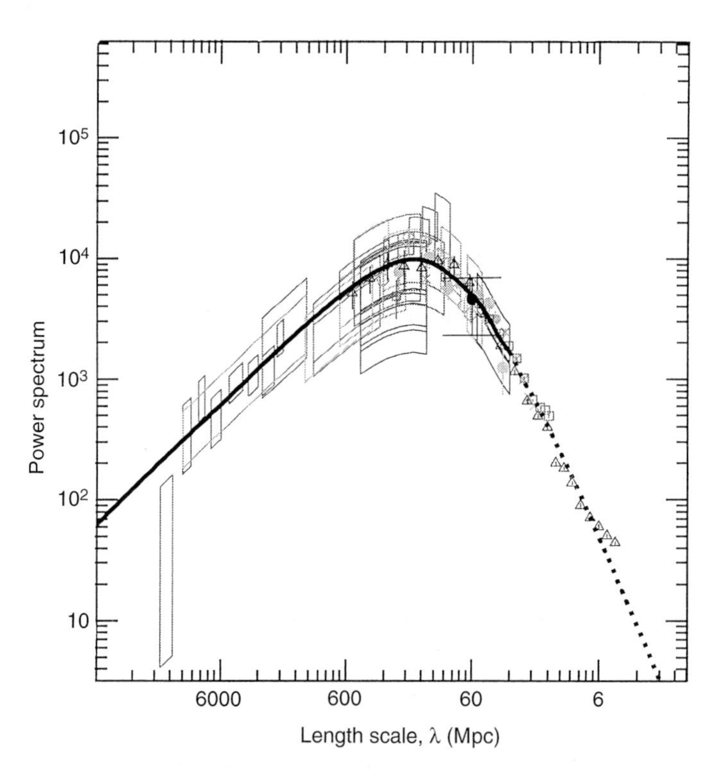


**Fig. 7.2** Power spectrum of density fluctuations on different scales, λ (measured in megaparsecs). The power spectrum measures the strength of the density fluctuations on different scales today. The predictions of the mixed dark matter model (thick, dotted line) is shown compared with observations of galaxies (triangles, squares, and circles) and of the microwave background (rectangles—size indicates range of uncertainty of observation). (Figure from Eric Gawiser and Joseph Silk of the University of California, Berkeley, *Science*, **280**, 1405 (1998).)

Santa Cruz leading a sustained programme of investigation into this scenario.

There are in fact three types of neutrino known, one corresponding to each of the three light fundamental particles, or 'leptons' (from the Greek *leptos*, or light, in contrast to the baryons, from the Greek *barus*, or heavy). The known leptons are the electron, the muon, and the tau lepton. The first of these is familiar and a key building block of atoms. The muon and the tau lepton are generated when nuclei travelling close to the speed of light collide with each other, for example in particle accelerator or research nuclear reactor experiments. The role of the muon and tau lepton is still something of a mystery. 'Who ordered that?' the Nobel Prize-winning physicist Isidor Rabi is said to have remarked when he heard about the discovery of the muon. The neutrinos corresponding to the three leptons are called the electron neutrino, the muon neutrino, and the tau neutrino. The neutrinos detected from the sun are electron neutrinos and we can already set quite strong limits on the mass of this neutrino. Muon neutrinos are detected in experiments using research nuclear reactors and accelerators. The tau neutrino has not yet been detected but is presumed to exist by theoretical arguments based on symmetry. If the neutrinos have a non-zero mass, the simplest explanations would predict that the most massive would be the tau neutrino, so this is the best bet for a cosmologically important neutrino.

For the past decade there has been a major effort directed at trying to measure the mass of the neutrinos. This has to be done rather indirectly. One of the likely consequences of neutrinos having a non-zero mass is that they would undergo 'oscillations' in which they would change from one type to another. Neutrino oscillations have been suggested as the explanation for the lower than expected rate at which electron neutrinos reach the earth from the sun. The basic idea is to try to prove that one type of neutrino has changed into another type, which would confirm that the two types involved had a non-zero mass. In one type of experiment trying to detect the transformation of a muon neutrino into a tau neutrino, a beam of muons is generated in an accelerator or nuclear reactor. The muons are directed at a target, where they generate a muon neutrino when they collide with a nucleus. This can not itself be seen but the idea is to look nearby for the track of a tau lepton, which would prove that the muon neutrino

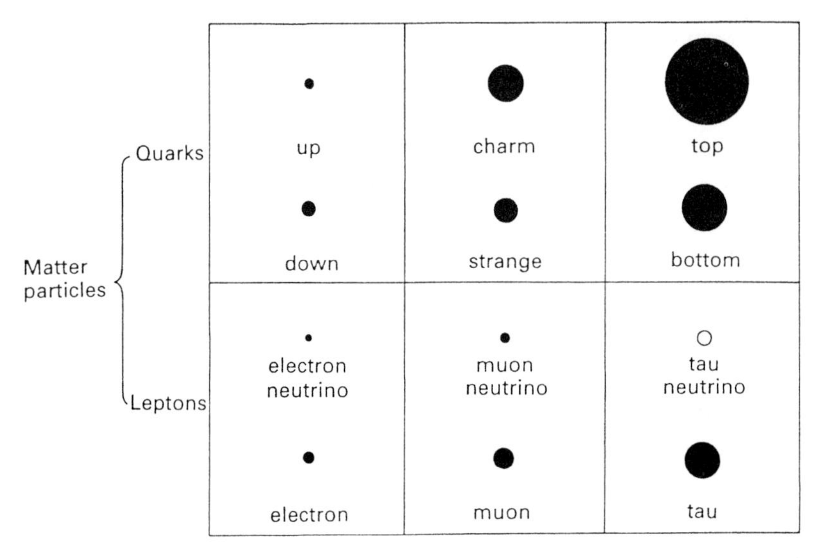


**Fig. 7.3** The three types of neutrino are associated with the three known types of lepton, or light particle: the electron, the muon, and the tau. In current particle physics theory, each of these leptons is grouped in a family with two quarks. The six quarks are building blocks of the baryons (neutrons, protons, etc.). The relative sizes of the particles in each row indicate the relative masses of the particles.

had changed into a tau neutrino which in turn caused the ejection of the tau lepton from another nucleus.

The second type of experiment involves careful study of the neutrinos that are part of the shower of secondary particles generated when an energetic cosmic ray collides with a molecule of air high in the earth's atmosphere. This shower of secondary particles can be studied at ground level and the flux of different types of neutrino measured. Recently a collaboration of 300 scientists, working at the Super-Kamiokande neutrino experiment located to the west of Tokyo, announced that they had detected neutrino oscillations in atmospheric muon neutrinos. The Super-Kamiokande experiment is a Japanese–American venture consisting of a tank of very pure water the size of St Paul's cathedral installed in a deep zinc mine 1 mile inside a mountain. As the neutrinos travel through the tank at close to the speed of light they cause tiny flashes of light to be emitted which are detected by a battery of very sensitive detectors. For some time physicists studying neutrinos in the earth's atmosphere have found that the proportion of muon neutrinos is smaller than predicted by theory. The main new evidence provided by the Super-Kamiokande results is that the proportion of muon neutrinos per second hitting their experiment from above was greater than the proportion hitting from below. The latter have traversed the whole earth before reaching the experiment and so have had more time to oscillate, that is change into another neutrino type. The team conclude that the difference in mass between the muon neutrino and, presumably, the tau neutrino is about one ten millionth of the mass of the electron.

In the simplest particle physics explanations of non-zero neutrino masses, where the mass of the tau neutrino would be much greater than the mass of the muon neutrino, the mass of the tau neutrino would then be about a ten millionth of the mass of the electron. This would be too low to make a significant impact on the density of the universe or the origin of structure. There are alternative theories in which the tau and muon neutrinos would have rather similar masses and in such models the mixed dark matter model would still be viable. It is certainly exciting that neutrinos have been proved to have a non-zero mass since this requires an extension of the so-called standard model of particle physics.

## The total density of the universe

The critical test between these two explanations, a non-zero cosmological constant and mixed dark matter, is the value of the density parameter $\Omega_{tot}$ (see p. 57). We can try to build up a census of where the matter in the universe is. If we take the visible parts of galaxies, the discs and bulges of spirals and the spheroidal components of elliptical galaxies, we arrive at a total contribution to $\Omega_{tot}$ of 0.006. Galaxies then have about ten times as much mass as this in the form of dark haloes. In our Galaxy we find that some of this dark halo, perhaps 20%, may be due to dark objects of stellar mass, probably white dwarfs, and this would account for double the mass of the disc. Most elliptical galaxies are found in rich clusters and these tend to have a cloud of hot gas which also amounts to about double the mass in galaxies. If we apply this factor to all galaxies, the contribution of baryonic matter in galaxies is increased to 0.018. We saw in Chapter 5 that the total contribution to $\Omega_{tot}$ from baryonic matter, as deduced from the nucleosynthesis of the light elements in the early universe, is 0.03. Thus about 40% of the baryons must lie outside galaxies (and clusters) and is presumably in the form of intergalactic clouds of hydrogen (and helium) which have not formed into galaxies. Such clouds are indeed detected in the line of sight to distant quasars through their absorption at ultraviolet wavelengths. The 80% (or more) of dark haloes of galaxies not accounted for by baryons contributes a further 0.05 to $\Omega_{tot}$, which brings the total so far to 0.08. What remains unknown so far is the contribution to $\Omega_{tot}$ from cold dark matter distributed rather smoothly outside galaxies and clusters and the contribution from other ingredients like hot dark matter.

Do we have any direct ways of estimating $\Omega_{tot}$? One method which depends on assuming that matter in galaxies is distributed broadly similarly to the total matter is to study how fast galaxies are moving around at random in the universe, their so-called **peculiar** motions (peculiar in the sense of individual, not odd). One galaxy whose peculiar velocity is particularly well determined is our own, through the **dipole anisotropy** of the microwave background radiation (see p. 32). We know that our Galaxy is moving with a velocity of about 600 km $s^{-1}$ towards the constellation of Hydra. We can use the IRAS galaxy surveys, which sample a large volume of the universe and cover most of the sky, to estimate what net attraction the nearby galaxy clusters and concentrations would exert on us. The direction agrees well with the direction we are travelling in and to get the required velocity we then have to assume that $\Omega_{tot}$ is 0.7 with an uncertainty of $\pm 0.2$. A value for $\Omega_{tot}$ of 0.1 is most unlikely, but values anywhere between 0.3 and 1 are permitted.

We can also perform this test by studying the large numbers of galaxies for which direct estimates of distances have been made, for example via the observed correlations between the typical rotation velocities of spirals and the random velocities in ellipticals, and the total luminosity of the galaxies. For these galaxies we can measure the redshifts and hence deduce what the peculiar velocities of the galaxies are, in the radial direction at least. These can then be compared with what would be expected from three-dimensional maps of the galaxy density distribution (derived from the IRAS galaxy redshift surveys) and hence $\Omega_{tot}$ estimated. The results are very similar to those derived above from the peculiar velocity of our Galaxy. For example, Avishai Dekel, of the Hebrew University, Jerusalem, and his collaborators recently found $\Omega_{tot} = 0.8 \pm 0.2$ from a study of this kind.

It seems that when we look at the dynamics of galaxies on large scales we find that $\Omega_{tot}$ is close to the critical value of 1. The dynamics of galaxies on large scales are closely related to how the galaxies are distributed, since it is the clusters and voids which give the galaxies their peculiar velocities. When studies of the large-scale clustering of galaxies are combined with the evidence from the ripples in the microwave background radiation, $\Omega_{tot} = 1$ models are again preferred. In a recent study using all the available evidence, Joseph Silk, of the University of California, Berkeley, and his student Eric Gawiser found that only the mixed dark matter model with $\Omega_{tot} = 1$ could fit all the data, from a wide range of possible cosmological scenarios.

Although I personally attach a lot of weight to these dynamical and large-scale structure arguments, it would be wrong to give the impression that cosmologists are agreed that $\Omega_{tot} = 1$. Quite the contrary, there is at the moment a strong current of interest in and support for models with a lower value of $\Omega_{tot}$, say 0.3. Several of the arguments for low total density of matter relate to clusters of galaxies. The first arises from the evolution of the number of clusters per unit volume as we look back in time. In a high density universe, the density fluctuations start growing at a relatively recent time, so we would expect to see few rich clusters of galaxies at redshift > 1. In a low-density universe the growth happens earlier, so we do not expect such strong evolution in the space density of clusters. Neta Bahcall and colleagues from Princeton find $\Omega_{tot} = 0.3 \pm 0.1$ from optical surveys for clusters. Carlos Frenk, from Durham University, and his colleagues find $\Omega_{tot} = 0.45 \pm 0.2$ from a sample of clusters selected through their strong X-ray emission. On the other hand Chris Collins and Doug Burke, from Liverpool John Moores University, recently reported $\Omega_{tot} = 1$ from a large X-ray selected cluster survey.

A second argument based on clusters is one that we have already mentioned (p. 96). Direct comparisons of the total mass in clusters in the form of galaxies and of hot X-ray-emitting intergalactic gas compared with the total dynamical masses of the clusters suggest that $\Omega_{tot}$ can not be higher than 0.6 (assuming the proportion of baryonic matter in clusters is representative of the universe as a whole).

Even more compelling than these cluster arguments for low $\Omega_{tot}$ is evidence from two large programmes searching for Type Ia supernovae (see p. 49) at high redshifts. There has been a long, and to date unsuccessful, history of attempts to measure cosmological parameters by observing particular classes of astronomical object at different redshifts. These traditional tests for cosmological parameters tried to find a 'standard candle',* some class of astronomical object which would look the same at all distances. By observing such objects at very large distances, with redshifts up to 1, say, we should start to see the effects of the geometry of the universe, which in turn depends on the value of $\Omega_{tot}$ (if there was a non-zero cosmological constant—see Chapter 8—this would also affect the apparent brightness of distant sources as a function of redshift). To date the only objects which could be used were some kind of galaxy, for example the brightest galaxies in clusters, and the problem was that we could not be sure how these have evolved in the past and to what extent merging together of pairs of galaxies is a major factor in their development. Recently an interesting development has been the detection of Type Ia supernovae at large distances. Over 100 supernovae are now known with redshifts in the range 0.3 to 0.8. If we could be sure that supernovae at these distances are identical to the ones we have studied locally, then we would have a powerful way to estimate the density parameter $\Omega_{tot}$ and the cosmological constant $\Lambda$ (see next chapter). Preliminary results seem to favour a low value of $\Omega_{tot}$, in the range 0.1–0.4. So the estimates of $\Omega_{tot}$ from the large-scale dynamics of galaxies and estimates from the geometry of the universe seem to be in conflict with each other. If the large-scale structure and dynamics arguments are correct, then the mixed dark matter model with $\Omega_{tot} = 1$ is the best bet at the moment. If the cluster arguments and supernovae estimates are correct, we have to abandon the simple Einstein–de Sitter model of the universe, for which $\Omega_{tot} = 1$, and can probably not invoke hot dark matter to solve our problems with the spectrum of density fluctuations.

## Tilted cold dark matter

We shall address another of the options, a non-zero cosmological constant, in the next chapter. However, we should anyway look again at our basic assumptions about how galaxy formation gets started. We

* The metaphor of the 'standard candle', used in cosmology to denote a type of star or galaxy which is supposed to have the same intrinsic luminosity, or power output, wherever it is found in the universe, goes back to the days before the invention of electric lamps when the international standard of luminosity, the lumen, was based on a carefully manufactured 'standard' wax candle.

saw that the different dark matter scenarios tend to start from the assumption that there was a very simple spectrum of density fluctuations laid down in the very early universe, probably at the inflationary epoch. Just what do we really expect from inflationary models? Inflationary theorists have been playing with a large zoo of models and have come up with a bewildering array of possibilities. Although these different possibilities tend to give density fluctuation spectra similar to the simple one we assumed earlier in this chapter, most tend to predict deviations from this simple form. In fact, whereas when the COBE fluctuations were announced Steven Hawking was quick to claim that it proved inflation had occurred, theorists are now hoping that astronomers will be able to deduce from the galaxy distribution and from microwave background fluctuations what the primordial density fluctuation spectrum was like, so that they can then decide which inflationary model is correct.

From the evidence discussed earlier in this chapter we saw that a simple cold dark matter picture with a Harrison–Zeldovich primordial density fluctuation spectrum (p. 93) does not work. If we fit it to the amount of structure we see on the scale of galaxies, the model does not predict enough power on large scales. We have discussed the possibility that some additional hot dark matter could have affected the evolution of the spectrum of density fluctuations in the required sense. However, another way out would be to stay with a purely cold dark matter scenario but abandon the Harrison–Zeldovich assumption. What we would need is a version of inflation which generates a spectrum of density fluctuations with a bit more power on large scales than the simplest Harrison–Zeldovich form. This would be characterized by the degree to which the spectrum of density fluctuations was 'tilted' towards larger scales (how $\Delta\rho/\rho$ differs from the inverse square dependence on scale mentioned above). So a 'tilted' cold dark matter scenario is another way of generating the required extra structure on large scales. We are just saying that, because of the particular form of inflation, there was more structure generated in the early universe on large scales than our simplest models predict.

## Making galaxies from topological defects

Another completely different idea for generating the density fluctuations which are needed to form galaxies is, as we saw in

Chapter 5, that 'topological defects' could play a role. When phase transitions occur in the early universe, for example the transition associated with the breaking of the Grand Unified Force, it is expected that defects could appear throughout space rather like the defects that appear in a block of ice when it freezes. In the cosmological situation these would be localized regions of very high energy density which could be point-like (monopoles), line-like (cosmic strings), sheet-like (domain walls), or have three-dimensional structure (texture). Cosmic strings are especially suitable for generating galaxies. They would form a spaghetti-like network permeating the universe and as the strings moved around, they could trigger the formation of galaxies in their wake. No evidence for strings has been found in the region of the universe that has been surveyed to date and the view is that the strings would necessarily have decayed away by the present time. But they may have left a signature on the density fluctuation spectrum which could be detected either in the galaxy distribution or in the microwave background fluctuations. The main effect is that the distribution of fluctuations on the sky would no longer look random. The importance of cosmic strings can be characterized by the average tension in the strings, denoted by $\mu$.

Our seventh cosmic number therefore characterizes the additional ingredient needed to make galaxies, either the density in hot dark matter $\Omega_{\mathrm{hdm}}$, the tilt of the primordial density fluctuation spectrum, or the cosmic string tension $\mu$. The total density of the universe $\Omega_{\mathrm{tot}}$ is not a new cosmic number because it can be written as the sum of the contributions from baryons (ordinary matter), cold dark matter and, if it exists, hot dark matter:

$$\Omega_{\mathrm{tot}} = \Omega_{\mathrm{b}} + \Omega_{\mathrm{cdm}} + \Omega_{\mathrm{hdm}}.$$

The value of $\Omega_{\mathrm{tot}}$ almost certainly lies between 0.1 and 1 and there are strong advocates for both ends of the range. There is substantial support among cosmologists at the moment for a value around 0.3, but I very much doubt that we have heard the end of this story. The seventh cosmic number, together with the cosmological constant which is the subject of the next chapter, remains the most uncertain of our nine cosmic numbers.

## Chapter 8

# How heavy is the vacuum?

*My greatest mistake*

Albert Einstein

The eighth number of the cosmos, the cosmological constant, Λ (lambda), was introduced by Einstein in 1917 in order to permit a static model of the universe. The idea is that a repulsive force, whose strength increases with distance, operates on the cosmological scale and prevents the universe from falling together under gravity. Recall (p. 25) that Newton, in his discussions with Richard Bentley, could not decide what the fate of a uniform, smooth, infinite universe would be, and argued that as the matter would not know which way to fall it would remain in its place. Quite apart from the question of the stability of such an infinite, static Newtonian universe, the issue is settled in Einstein's general theory of relativity. An infinite, uniform universe could not be static without the addition of the cosmological repulsion.

As de Sitter, Lemaître, Eddington, and Friedman showed that expanding models without a cosmological constant were possible and as Hubble showed that the universe was indeed expanding, Einstein regretted the unnecessary complication of the cosmological constant and called it 'my greatest mistake'. He felt that the elegance and simplicity of general relativity had been compromised by the introduction of the additional, and in his view unnecessary, term. Since then fashions among cosmologists have oscillated between believing the cosmological constant should be zero and believing that a non-zero cosmological constant may play a crucial role in understanding the history and geometry of the universe.

## Models with non-zero cosmological constant

The cosmological repulsion offers some universe histories dramatically different from the standard model, in which all solutions expand from a Big Bang and the future offers only two choices: expansion for ever or recollapse to a Big Crunch. Einstein introduced the $\Lambda$-term in order to allow the possibility of a static universe. Although this does not correspond to reality, and anyway such a universe is unstable, there is another possible model found by Eddington and Lemaître in which the universe expands from a static state at an infinite time in the past. Another very different possibility is a contracting universe in which the cosmological repulsion is able to halt the contraction and turn it into expansion, the 'bounce' model. In a class of model studied by Lemaître in 1929 the universe expands from a Big Bang and gravity almost succeeds in halting the expansion but the cosmological repulsion eventually wins and the expansion continues at an ever-increasing rate. In the Lemaître models the universe has a long 'coasting' phase in which it is almost static, so the age of the universe can be very long, much longer than the Hubble time. Such models were reexamined independently in the late 1960s by the Russian cosmologist Nikolai Kardashev and by myself to see whether the coasting phase could explain a claimed accumulation of quasar redshifts at around a redshift of 1.95. The British astrophysicist Geoffrey Burbidge argued strongly at the time that this apparent peak in the redshift distribution must have some theoretical significance. However, the peak disappeared when larger samples of quasars became available, so the Lemaître models were no longer needed.

One amusing aspect of the Lemaître models is that there is the possibility that the time spent in the coasting phase is long enough to allow light to completely circumnavigate the universe if the latter is finite. The Lemaître models are all universes with positive spatial curvature and in such models the universe could be of finite size. After travelling in a straight line in one direction away from earth you could eventually find yourself passing the antipodes or 'antipole' of the universe and heading back towards the earth from the opposite direction. You would not come across an edge to the universe and at every point on your journey you would still seem to be at the centre of an isotropic, expanding universe. It is analogous to travelling in two dimensions on the surface of a sphere. If you are sailing round the world you do not come to an edge anywhere and there is nothing special about the antipodes of your starting point. In the case of the universe we do not know whether you would actually find yourself heading back to earth. You could equally well find that you had headed off into a different universe, since we do not know how the universe connects together on large scales. If there was enough time to see round to the antipole, though, we would find that galaxies near there would seem very bright and large to us here. Since we have not found any large bright galaxies with high redshift, presumably we do not live in a universe where we can see to the antipole.

Even if we do not live in one of these more extreme cosmological constant universes, the $\Lambda$-term could have one important effect on the universe. The repulsion always has the effect of slowing down the deceleration of the expansion due to gravity and as a result these models always have longer ages than corresponding models with zero cosmological constant and the same matter density. This could be important because, as we saw in Chapter 4, the standard models may have trouble reconciling the age of the universe deduced from the Hubble constant and the age of our Galaxy derived from the stars in globular clusters. Observational cosmologists have therefore always wanted to keep the possibility of a non-zero cosmological constant. As usual we measure the cosmological constant in dimensionless form. The cosmological constant introduced by Einstein into his field equations has the dimensions $1/(\text{time})^2$, so we rescale it by dividing by the square of the Hubble constant, to make $\Lambda$ dimensionless. To give an example let us suppose that we know that the density parameter (p. 56), $\Omega_{tot}$, is 0.1. Then, with zero cosmological constant, the ratio of the age of the universe to the Hubble time would be 0.89, whereas if $\Lambda$ was 0.9, the ratio would be 1.28. For our preferred Hubble constant of 65 km $s^{-1}$ $Mpc^{-1}$ (Chapter 3), for which the Hubble time would be 15.1 billion years, the corresponding ages would be 13.6 and 19.3 billion years. So the cosmological constant can make a big difference to the age of the universe.

Supporters of the inflationary universe idea have another reason for wanting the possibility of a non-zero cosmological constant. They

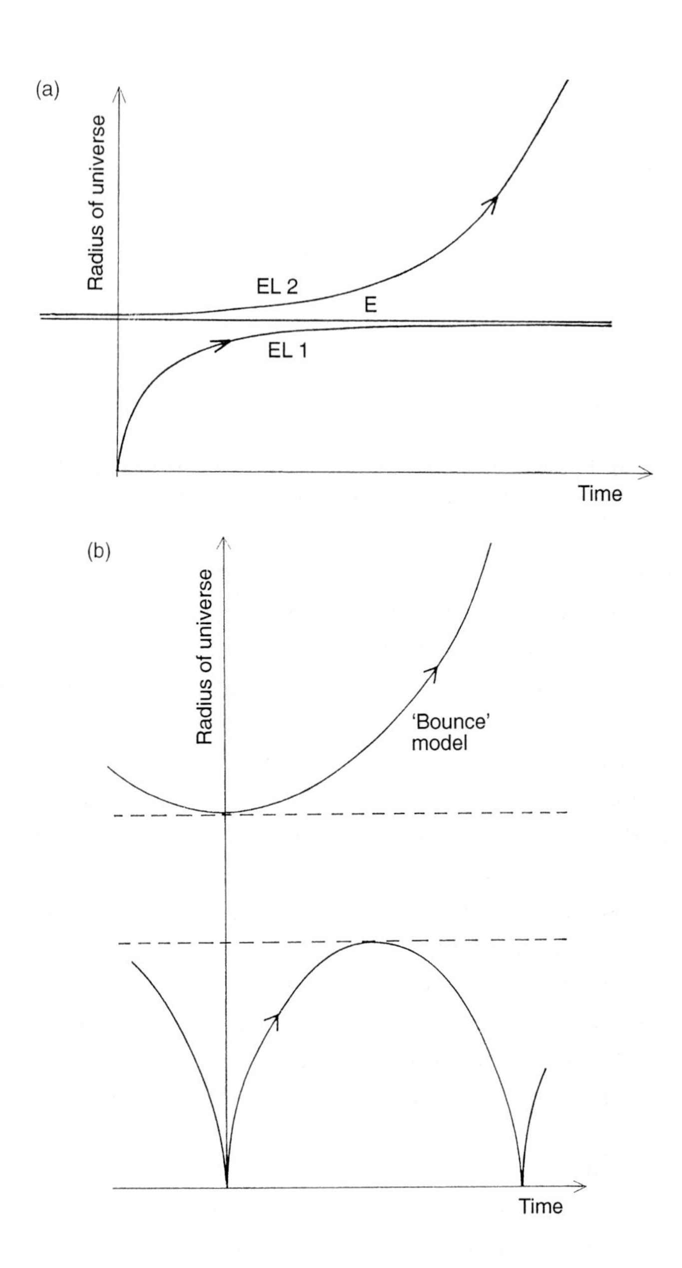
(a)
Radius of universe
EL 2
E
EL 1
Time
(b)
Radius of universe
'Bounce'
model
Time

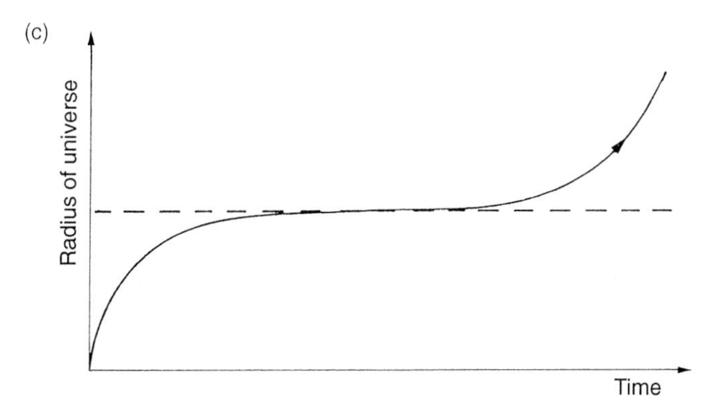


**Fig. 8.1** The history of a universe with cosmological repulsion $\Lambda$. (a) The Einstein static model (E) and the Eddington–Lemaître models (EL1, EL2). (b) The 'bounce' model and an oscillating model. (c) A Lemaître model, with long 'coasting' phase.

argue that inflation has the effect of driving the spatial curvature close to zero. The analogy is measuring the curvature on the surface of a balloon as it is blown ever larger. As the balloon expands, the curvature of its surface gets smaller. So even if the universe was strongly spatially curved before inflation started, by the time inflation ends the universe would be extremely flat and would probably still be flat today. For a zero cosmological constant, the model with zero curvature is the Einstein–de Sitter critical density model, so we would probably be close to this model if there had been inflation. This model makes very specific predictions about the density of the universe ($\Omega_{tot} = 1$) and the age of the universe (two thirds of the Hubble time), both of which may be problematic from an observational point of view. With a non-zero cosmological constant, the condition for zero curvature becomes $\Omega_{tot} + \Lambda = 1$, so we can satisfy this for example if $\Omega_{tot} = 0.1$, and $\Lambda = 0.9$, or if $\Omega_{tot} = 0.3$, and $\Lambda = 0.7$.

## The meaning of $\Lambda$

To the particle physicists the cosmological constant has a very different meaning. If we look at the equations of motion for the universe, then the $\Lambda$-term continues to operate when $\Omega_{tot}$ is zero so it is a property of the vacuum. In fact the cosmological constant has the same form in the equations of motion as the density, so it behaves like the energy density of the vacuum. The cosmological constant tells us how heavy the vacuum is. It may seem strange to think of the vacuum as possessing energy but in modern particle physics the vacuum is a seething mass of pairs of particles and antiparticles which come into existence for a fleeting instant and then annihilate, so-called 'virtual' particle pairs. When particle physicists estimate what is a 'natural' value for the energy density of the vacuum, they come up with a value equivalent to $\Lambda = 10^{120}$, 1 followed by 120 zeros, a mind-boggling number. In the actual universe it is clear that $\Lambda$ could be about 1 but it certainly is not as big as 10, let alone $10^{120}$. At present, particle physicists can not explain why the observed $\Lambda$ is so small. So they tend to argue that there must be some fundamental principle which they have not yet discovered that forces the cosmological constant to be exactly zero in the present-day universe.

On the other hand the whole basis of inflation is that following a phase transition the universe finds itself in a state where the vacuum really does have this enormous energy density. As a result the universe is driven to expand at an exponentially increasing rate. After a while the energy of the vacuum is converted into radiation and matter, and the universe continues with a normal expansion, with the cosmological constant now back to zero. As we saw in Chapter 5, this inflationary phase solves several philosophical problems in cosmology, the most important of which is the horizon problem.

In the last chapter we discussed what additional ingredient might be needed to bring the cold dark matter scenario into agreement with the spectrum of density fluctuations deduced from studies of the large-scale clustering of galaxies and the anisotropies in the microwave background radiation. We came up with three alternative ways of doing this: tilt, strings, or hot dark matter. A further way of resolving this problem is to have a non-zero cosmological constant and this idea has several strong supporters. Models with $\Omega_{tot}$ in the range 0.1 to 0.3 and with $\Lambda$ chosen to bring $(\Omega_{tot} + \Lambda)$ exactly to 1 so that the universe will be spatially flat, as predicted by inflationary theory, give quite a good fit to the observations (though not as good as the mixed dark matter model according to the study by Gawiser and Silk discussed on p. 113).

## Observational limits on the cosmological constant

Neither particle physicists nor relativists particularly like the idea of a relatively small non-zero cosmological constant. Can we set any limits from observations? The best direct limit has come from studies of the statistics of gravitationally lensed quasars. The first example of a gravitationally lensed system, 0957+561, was discovered by radio-astronomers in 1979 (see p. 53). The system appears to consist of two very similar quasars separated by about 6 arc seconds. In fact the two quasars are both images of the same quasar, the light having been bent by two different routes around an intervening galaxy. Careful image processing reveals the intervening galaxy close to one of the quasar images. The intervening galaxy acts like a lens to magnify and distort the image of the background source. If the alignment of lensing galaxy and background source is perfect a ring of light is seen, known as an 'Einstein ring'. More usually there are two to four images seen, often in the form of bright arcs. Dozens of other examples are now known, with the background source usually a quasar, sometimes a galaxy. Beautiful examples of lensed arcs are seen against rich clusters of galaxies, with the whole cluster acting as a lens and the arcs being lensed images of background galaxies. Surveys for lensed quasars conducted at radio wavelengths allow statistics to be built up about the numbers and redshifts of the lensed quasars. These depend rather sensitively on the geometry of the universe and in particular on the value of the cosmological constant. Current statistics suggest that $\Lambda$

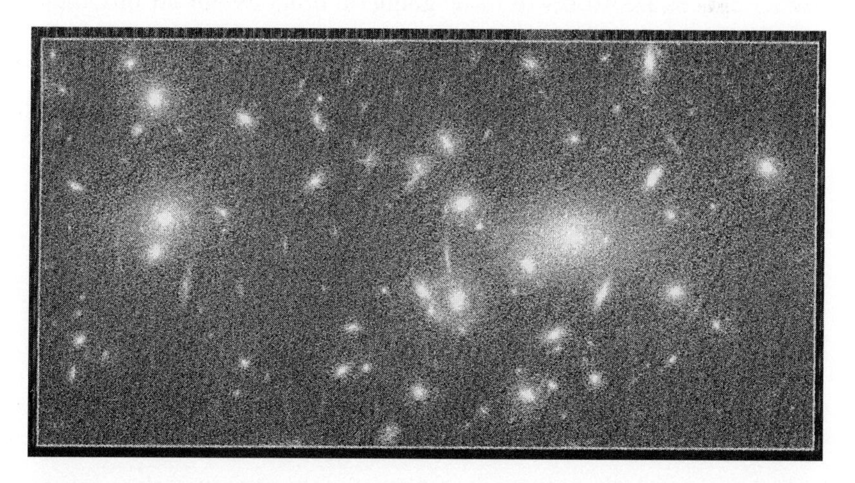

**Fig. 8.2** Gravitational lens arcs in the cluster Abell 2218, from an image taken with the Hubble Space Telescope.

can be no larger than about 0.9, but it is not yet possible to rule out the range of particular interest, 0.7–0.9.

We saw in Chapter 4 that the current best estimates of the age of the universe and of the Hubble constant do not suggest any very severe age problem for the simple $\Omega_{tot} = 1$ Einstein–de Sitter model. If anything it is the low $\Omega_0$, $\Lambda > 0$ models which have an age problem in the opposite direction—they predict too long ages. For a Hubble constant of 65 km $s^{-1}$ $Mpc^{-1}$, the corresponding age of the universe for models with $\Lambda = 0.7$–$0.9$ and $(\Omega_{tot} + \Lambda) = 1$ would be 14.5–19.3 billion years. For $\Lambda = 0.7$ this is just about acceptable but for $\Lambda = 0.9$ this would definitely be too high compared with the recent estimates we discussed in Chapter 4.

Recently a new line of observational evidence about the cosmological constant has appeared, from studies of distant supernovae. For several years two teams, a US–British collaboration led by Saul Perlmutter of Berkeley, and a team led by Brian Schmidt of Mt Stromlo, have been searching for supernovae in distant galaxies, using regular observations with 4-metre telescopes to find the supernovae and the Keck 10-metre telescopes to measure the redshifts. They have now accumulated between them over 100 supernovae in galaxies with redshifts in the range 0.3–0.8. Assuming that these are all normal Type Ia supernovae (p. 49) and that such supernovae at these earlier epochs are similar to those going off today in our locality, these supernovae can be used to try to estimate cosmological parameters like $\Omega_{tot}$ and $\Lambda$ from the dependence of the apparent brightness of the supernovae at maximum light on redshift. Saul Perlmutter and his colleagues claim that their study shows that $\Omega_{tot}$ must be 0.2–0.4 and that $\Lambda$ must be $> 0.5$. This is a dramatic claim and will have an enormous impact on cosmology if correct. The second collaboration led by Brian Schmidt is also reaching similar conclusions. However, there are some doubts about whether exactly the same kinds of supernovae are being found at large distances as are found locally. Even with nearby Type Ia supernovae we do not know for certain the mechanism that leads to the explosion, whether it is a single white dwarf star on which gas has been dumped from a companion or whether it might be a merger of two white dwarfs in a close binary system. There is also a possibility that the effect of extinction by dust

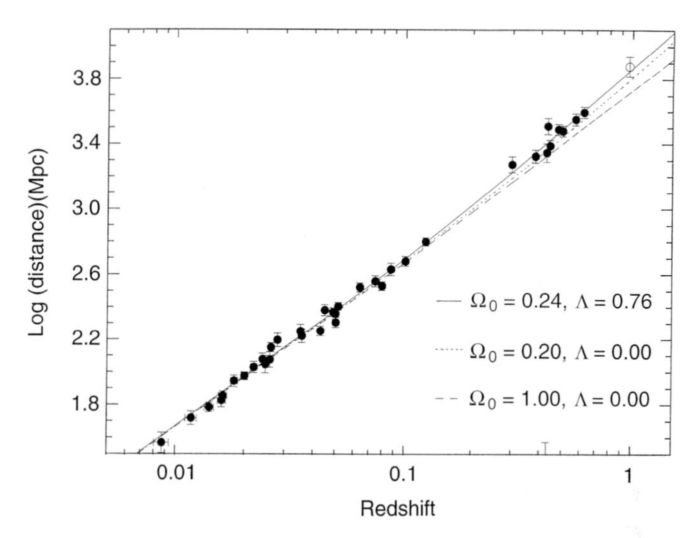


**Fig. 8.3** Hubble diagram for high redshift supernovae. The broken line is the model with $\Omega_0 = 1$, $\Lambda = 0$ and does not seem to be a very good fit. The best model is the one with $\Omega_0 = 0.24$, $\Lambda = 0.76$ (solid line). (Figure from the consortium led by Brian Schmidt of Mt Stromlo Observatory.)

in the galaxies may have been underestimated. My feeling is that the precision of the method has been overstated and that the model with the critical density $\Omega_{tot} = 1$ can not really be ruled out yet with confidence. But it is extremely impressive that these distant supernovae can now be routinely found and studied and this is bound to have huge ramifications.

We saw that cosmologists prefer models with zero spatial curvature because of the predictions of inflation, so they expect to find $\Omega_{tot} + \Lambda = 1$. Post-COBE observations of the ripples in the microwave background radiation, made either from the ground or from balloon-borne telescopes, which are sensitive to the value of $(\Omega_{tot} + \Lambda)$, have supported a value of about 1, though still with wide uncertainties. Because there are some lines of evidence that point to values of $\Omega_{tot}$ significantly less than 1, say in the range 0.1–0.3, cosmologists have been perhaps rather quick to accept a compensating positive value for the cosmological constant. In Chapter 7 we saw that while there are some arguments favouring a value of $\Omega_{tot}$ around 0.3, there are other equally compelling arguments for $\Omega_{tot} = 1$.

When inflation first appeared on the scene it was attractive not only because it solved the horizon problem and offered an origin for density fluctuations but also because it seemed to make very definite predictions about $\Omega_{tot}$ and the spectrum of the initial density fluctuations. However, further detailed studies have changed the picture. Firstly it is clear that, far from predicting the initial spectrum of density fluctuations, things are the other way around. We have to determine the initial spectrum of density fluctuations by other means, for example by studying the microwave background fluctuations on small angular scales, in order to find out the detailed physics of the inflation process. Even a spatially flat universe ($\Omega_{tot} + \Lambda = 1$) is not actually a prediction of inflation. How flat the universe ends up being today depends on how curved the universe was when inflation started and how long inflation goes on for. A spatially flat universe might be taken as evidence that inflation may have occurred in the past but the reverse does not hold. Inflation has in my opinion become a much less important and attractive idea because it does not in fact make any concrete predictions about the universe.

We cling to inflation because we can not see any other way of solving the horizon problem except by 'initial conditions', which in the cosmological context means phenomena at the Planck time. But in our present state of ignorance about what happens at or before the Planck time, contrary to what you might have been led to believe by some of the more over-enthusiastic theorists, there is no good reason for supposing that the solution of the horizon problem and of the origin of density fluctuations does not lie at that epoch. There is no good reason for believing in inflationary theory unless it makes testable predictions about the universe.

On balance I remain sceptical about the cosmological repulsion and suspect that in the present-day universe the cosmological constant is close to zero. Hopefully we can make progress in the next few years in determining the cosmological density parameter $\Omega_{tot}$ more precisely. Most current observations would be consistent with a universe with cosmological constant $\Lambda = 0$, cosmological density parameter $\Omega_{tot} = 0.3$, Hubble constant $H_0 = 65$ km s$^{-1}$ Mpc$^{-1}$, age of the universe $t_0 =$ 12–13.5 billion years (and age of our Galaxy 11–12 billion years). A non-zero cosmological constant would be, if anything, an embarrassment, resulting in too great an age for the universe. The studies of large-scale velocity flows using IRAS galaxies and of large-scale structure in the galaxy distribution and in the microwave background radiation remain at odds with this synthesis, suggesting values for $\Omega_{tot}$ close to 1, and this would require the value for either the Hubble constant or the age of the universe to be at the low end of the permitted ranges. It will be exciting in the years ahead to see which of these pictures is correct.

Chapter 9

# How do galaxies form?

*A time to be born, and a time to die*

Ecclesiastes

We have seen how the shape of the distribution, or spectrum, of primordial density fluctuations over a wide range of scales is a powerful probe of the cosmological scenario and of the nature of the dark matter in the universe. However, even if we knew the form of this spectrum exactly we would still not know how galaxies form and evolve because this depends on the detailed physics of the formation of stars, how they interact with the surrounding interstellar gas, and how they lose mass at the end of their lives.

Almost all of the viable scenarios for the formation of structure in the universe assume a dominant role for cold dark matter. According to these scenarios, cold dark matter density perturbations of mass a few million solar masses would be quite well developed at the era of recombination and would start to cluster together to form the dark galaxy haloes. Fluctuations in the density of ordinary baryonic matter do not start to grow until the era of recombination (the epoch a few hundred thousand years after the Big Bang when the universe first becomes transparent to radiation; see p. 77), at which time the difference in density from the average would be less than 1 part in 10 000. At first a particular blob of matter would still be expanding with the universe but the expansion would now begin to slow down under the influence of the additional gravity of the blob. These protogalactic fragments would eventually reach a maximum size and then start to fall together as gravity gradually exerted its dominance over the expansion of the universe. As this was happening the fragments would also be feeling the pull of the nearest dark matter halo and starting to fall towards it. Eventually the protogalactic gas cloud would fall together at the centre of the dark matter halo and stars would form.

So our ninth cosmic number characterizes the history of galaxy and star formation. It turns out that it may be quite difficult to do this with a single number. We definitely need one number to define how fast the average rate of star formation in galaxies has been changing (declining) over the past few billion years. But, as we shall see, we may need a second number to define the epoch at which star formation reached its peak. The question of whether this takes us over our budget of nine numbers will be resolved in Chapter 10.

## How do galaxies form?

The first generation of stars would be those of the globular clusters in the haloes of galaxies. Soon after this, perhaps in the same phase, the stars of ellipticals and of the central stellar 'bulges' of spiral galaxies would form. And then 1 to 3 billion years later the first stars in the discs of spiral galaxies would form. But the details of this formation process remain an enigma. In fact, until recently almost no progress had been made at all on understanding exactly how and when galaxies form. All that could be said about the epoch at which galaxies formed, marked by the moment that stars start to shine in the galaxies, was that it must be at least 100 million years after the era of recombination and at least several billion years before the present. In terms of the redshift at which galaxies first appeared, this narrowed the range down to somewhere between 1 and 30. Until 1995 almost no galaxies had been found with redshifts greater than 1 unless they were exceptionally powerful emitters of radio or far infrared radiation (or contained a quasar—see below). And across the range of redshift and hence of look-back time over which galaxies could be studied little evidence could be seen that galaxies had changed much, so that the epoch of galaxy formation had to be set much further back in the past. Studies of the redshift distribution in moderately faint galaxy samples claimed to show that the luminosity or optical power output of the galaxies could have changed little from redshift 0.5 to the present.

## The evolving radio-galaxy and quasar population

However, populations of galaxy exist in which we have known for decades that there are very strong changes in the population with time. In the 1960s it had become clear that there was strong 'evolution' with cosmic epoch in the extragalactic radio-source population. As the sky was surveyed to deeper flux levels the number of radio sources per unit area on the sky increased faster than expected for a uniformly distributed population. At first, in the 1950s, there had been controversy between Cambridge and Australian radio-astronomers over how steep this increase was, but this was resolved by improved methods of analysing the data. There was also controversy about whether the sources were nearby stars in our Galaxy or much more distant galaxies. In 1963 quasi-stellar radio sources, or quasars, were discovered by Cyril Hazard and Maarten Schmidt. These turned out to be galaxies at high redshift with very powerful sources of optical and radio emission in their nuclei. On a photographic plate they look like stars because the nuclear source blots out the galaxy's surrounding starlight, hence the name. Quasars make up about a third of the brighter radio sources, away from the plane of our Galaxy where many bright radio sources associated with newly forming stars are found, while the remaining two thirds are luminous radio galaxies.

**Fig. 9.1** The quasar 3C273.

In 1966 the Cambridge radio-astronomer Malcolm Longair proposed that the steep radio source-counts were due to strong changes in the quasar population with epoch, with the number of quasars per unit volume increasing steeply as we look back to earlier times (after correction for the effects of the expansion of the universe). So a higher proportion of galaxies would contain quasars at earlier epochs. Two years later I showed first that the radio-galaxy population was also undergoing strong changes with epoch. I also found a better fit to the available data if it was the typical luminosity rather than the space density of the population that was increasing with redshift. The change with epoch, or 'evolution', was so strong that by redshift 2 the typical luminosity of the population has increased by a factor of 30 (or alternatively the space density has increased by 750). As we are looking back in time when we look out to higher redshift, we should really restate this by saying that since the epoch at which the redshift is 2 (when the universe was about 2 to 3 billion years old) the luminosity of the population has decreased by a factor of 30.

In the intervening 30 years the radio source-counts have been explored to flux levels a million times fainter than those used by Malcolm Longair and myself but the picture of strong evolution has remained intact, with the form of the evolution confirmed as broadly a shift in the characteristic luminosity of the population. Deep surveys for quasars at optical wavelengths have shown that optically selected

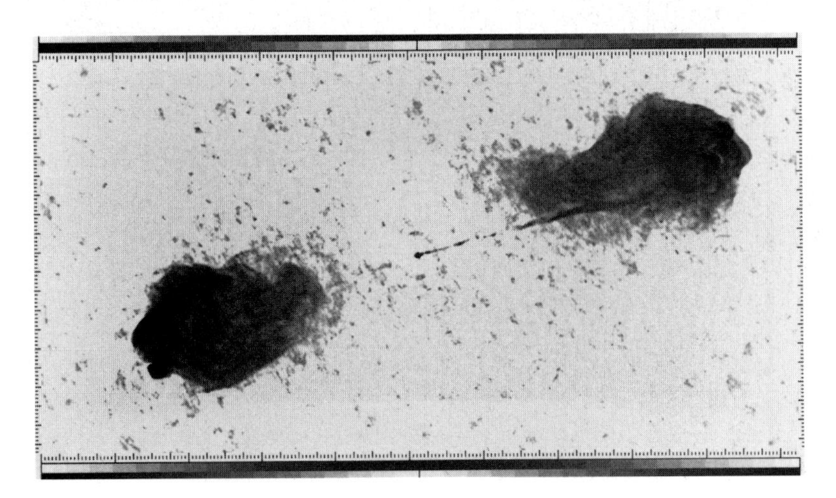

**Fig. 9.2** The radio galaxy Cyg A.

quasars, most of which are relatively weak radio emitters, also show the same strong evolution. The typical luminosity of the population increases steeply to redshift 2, has a plateau to redshift 3 or so, and then probably declines towards higher redshifts. So it seems that the quasar phase of galaxies reached a peak at the epochs between redshifts 2 and 3. What could this be due to? To make headway on this we probably need to turn to another type of galaxy which also shows signs of strong evolution with cosmic epoch.

## Starburst galaxies and their evolution

In 1983 the IRAS satellite burst on the world with its new picture of the sky at far infrared wavelengths (10–100 microns) and the concept of 'starburst' galaxy, which was soon found to be the dominant type of active galaxy. Even before the launch of IRAS it was clear from pioneering work from aircraft and balloons that there were galaxies that were exceptionally powerful at far infrared wavelengths. One example was the nearby galaxy Messier 82 (M82), a peculiar and dusty

**Fig. 9.3** The starburst galaxy M82.

galaxy in which an exceptional level of star formation is occurring. In the 1960s this had been interpreted as an exploding galaxy but this had turned out to be a mistaken analysis of light reflected from huge clouds of dust and gas associated with the galaxy. In a galaxy like our own about one third of the light from stars is absorbed by dust and reemitted at far infrared wavelengths. M82, on the other hand, emits ten times as much energy in the far infrared as is seen in optical starlight. The reason is that newly forming stars are still embedded in the clouds of dust and gas from which they are forming and almost all their light is absorbed by dust and reemitted at far infrared wavelengths.

The IRAS survey of the sky found tens of thousands of starburst galaxies—we define these to be galaxies in which the far infrared luminosity exceeds the optical luminosity, generally clear evidence of a strong burst of new star formation. I described in Chapter 7 how my collaborators and I set out to measure the redshifts of a large sample of these galaxies in order to map the galaxy distribution on the large scale. We were also able to use this survey to study the evolution of the starburst galaxy population. To our surprise we found that their characteristic luminosity or space density also changed with redshift, or look-back time, at about the same, very strong, rate as the quasars and radio galaxies. Subsequently we have been able to confirm this result with larger samples and deeper IRAS surveys. This strongly suggests that there is a link between the evolution of quasars and radio galaxies on the one hand and starbursts on the other.

## What is the link between ultraluminous infrared galaxies and quasars?

Another discovery from these IRAS redshift surveys was that there are large numbers of exceptionally luminous infrared galaxies, the 'ultraluminous' infrared galaxies, with luminosities greater than a million million times the total output of the sun. This luminosity puts these galaxies in the same power range as the quasars. However, in any volume of the universe there are ten times as many ultraluminous infrared galaxies as there are quasars at the same power output. There is still controversy about what powers these ultraluminous infrared galaxies. Dave Sanders, now at the University of Hawaii, and his colleagues believe that their huge power means that they must be powered by quasars and that what we are witnessing is the formation of quasars. On the other hand I and others have always argued that they are powered primarily by starbursts and are simply more extreme versions of the more prevalent starbursts like M82. Crucial light has been shed on this by a programme of work using the Infrared Space Observatory (ISO) led by Reinhard Genzel of the Max Planck Institute for Extraterrestrial Physics at Garching, near Munich. He and his team have used the short-wavelength spectrometer on ISO to study infrared emission lines from ultraluminous infrared galaxies in the wavelength range 3–40 microns. They discovered that in most of these

**Fig. 9.4** The Infrared Space Observatory (ISO) on the ground.

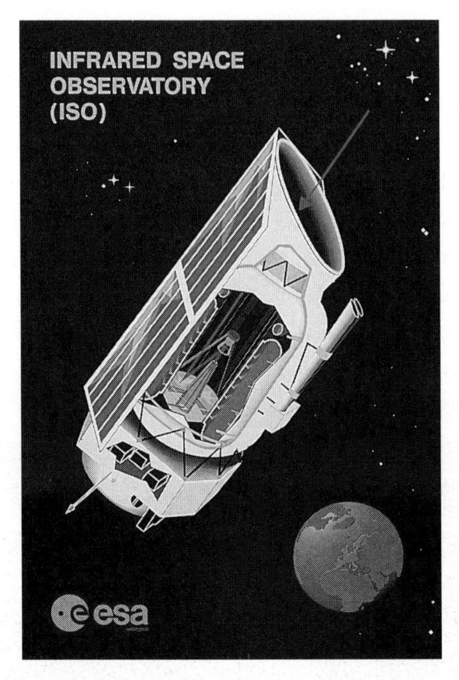


**Fig. 9.5** Artist's impression of ISO in orbit.

ultraluminous infrared galaxies the emission lines found were characteristic of the moderate excitation expected from hot stars rather than the very high excitation lines seen in quasars. This strongly suggests that ultraluminous infrared galaxies are indeed predominately powered by starbursts.

Now it is true that when we look at the very highest luminosity infrared galaxies, those with luminosities higher than 10 million million times the total output of the sun, we find that an increasing proportion of them contain active quasar-like nuclei. So there probably has to be a link between the powering of the far infrared output and of the active nucleus. We also find that as w[illegible]rds higher infrared luminosities there is increasing evide[illegible] galaxies are undergoing interactions or mergers with ot[illegible]. Can we put all this evidence together to provide a mechanism for the rapid rate of evolution that we see both in quasars and in starburst galaxies?

In radio galaxies and quasars we believe that the nuclear activity is powered by gas falling into a massive black hole, of mass perhaps a hundred million solar masses. It is known that galaxy–galaxy interactions and mergers have the effect of feeding gas and dust towards the nuclei of the galaxies. If a black hole were already present then we might expect that it would light up as a quasar (or radio-galaxy) event during a strong interaction. It is possible that most galaxy nuclei contain a black hole and the difference between active galactic nuclei containing quasars or radio galaxies and the rest is simply that the latter are not being fed at the moment.

Gas falling into the nucleus of a galaxy is also what is needed to trigger a massive starburst. So we see the possibility of a common mechanism, the fuelling of a galactic nucleus with gas due to a strong interaction or merger. And this process of galaxy merging is clearly an integral part of the galaxy formation and building process. Galaxies do

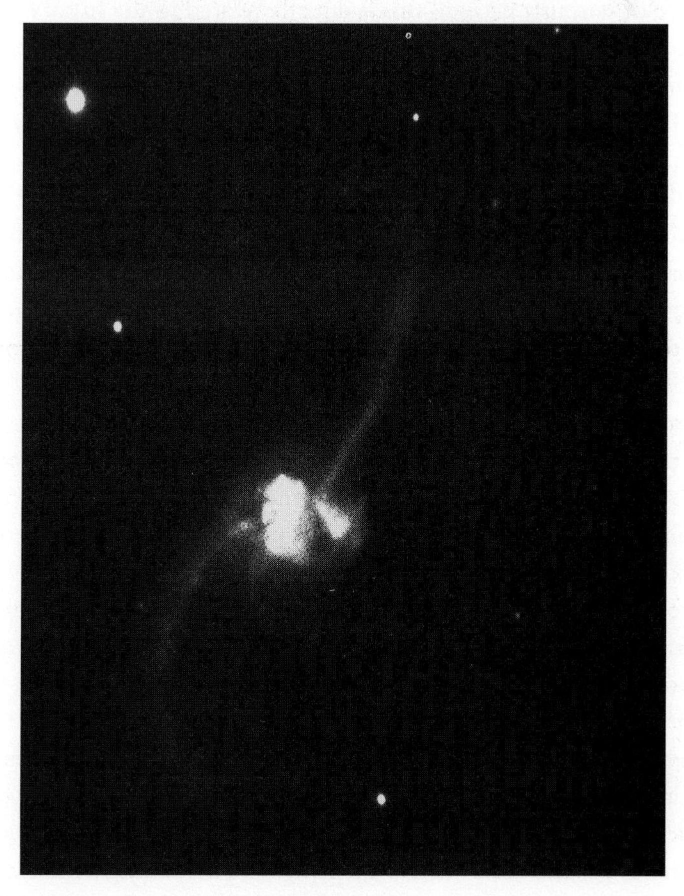

**Fig. 9.6** Merging galaxies: the 'Antennae'.

not just form as isolated objects, switch on their stars, and shine in glorious isolation. They are formed in groups and clusters and they continually interact with each other gravitationally, often merging with their near neighbours after a series of close encounters. A galaxy like ours might have been built up out of ten smaller galaxies, with each process of merging accompanied by an impressive starburst episode. Our Galaxy probably did not generate dramatic quasar events, though, because the black hole in the centre of our Galaxy is not especially large, only a few million solar masses.

So we should think of the strong evolution we see in quasars and in starburst galaxies as the history of the gas supply to galactic nuclei from interactions and mergers. The form of the evolution tells us that from redshift 2 to the present the gas supply has been steadily declining. In the case of starburst galaxies this is directly related to the history of the star formation rate. In the case of the quasars the gas supply controls the luminosity of the quasar events. As we push out to larger redshifts, back further in time, we might expect to see other factors playing a part in the process. Eventually we will see back to the era when the first galaxy fragments form and make stars. We must also eventually see an era when black holes start to grow in galactic nuclei. Enormous interest has therefore focused on detecting protogalaxies, galaxies in the first stages of formation.

## The search for protogalaxies

The search for galaxies in the process of formation has been a long and, until recently, unsuccessful story. The Princeton astronomers Bruce Partridge and Jim Peebles proposed in 1967 that forming galaxies would be bright extended sources at very high redshift (10–30), whose ultraviolet light would be redshifted into the near infrared band. Their idea was that the first major episode of star formation would occur when the protogalactic cloud was at its maximum extent. Later models suggested that the first stars would form when the protogalactic cloud collapsed together and predicted that protogalaxies would be very bright emitters of ultraviolet radiation, especially in the 'Lyman alpha' emission line of hydrogen at a wavelength of 0.1216 microns. Starting in the 1980s there were a series of searches for galaxies in which the Lyman alpha was highly redshifted, in the hope of picking up galaxies at such great distances, and hence so far back in time, that they would be in the process of formation. Until recently the only objects found were galaxies close to, and at the same redshift as, quasars and for these there was a suspicion that the Lyman alpha emission was being excited by ultraviolet radiation from the quasar rather than indicating a burst of star formation.

In 1991 my collaborators and I identified a faint IRAS 60-micron source with a very faint galaxy at a redshift of 2.3. The galaxy, which was known by its catalogue name, IRAS F10214+4724 (the numbers describe its position on the sky), had bright emission lines but it became clear that these lines were illuminated by a weak quasar in the galaxy's nucleus. More interestingly, the total luminosity in far infrared radiation was 300 million million solar luminosities, 30 000 times more than the whole output of our Galaxy. Could this be an example of a galaxy in formation? A large mass of dust and molecular gas was detected through observations at submillimetre wavelengths, confirming the starburst interpretation of the far infrared output. However, imaging with the Keck Telescope and with the Hubble Space Telescope showed that it was a gravitationally lensed galaxy (p. 50) and that the true far infrared luminosity of the galaxy was about a factor of ten lower after correction for the lensing magnification. This is still an exceptionally luminous galaxy, though. Over 30 galaxies are now known whose far infrared luminosities exceed 10 million million solar luminosities, with redshifts ranging from 1 to 4.8, and if, as seems likely, the far infrared emission is powered by a starburst then these are galaxies in the process of a very major episode of star formation, with stars forming at over a thousand solar masses per year, compared with about one solar mass per year in our Galaxy today.

## The breakthrough to the high redshift universe

In the last few years there has been a breakthrough in techniques to find high-redshift galaxies. Simon Lilly led his colleagues in the Canada–France Redshift Survey which studied hundreds of faint galaxies in the near infrared 'I' band at 0.9 microns, finding many galaxies in the redshift range 0.5–1. For the first time they found direct evidence of steep luminosity evolution in the optical galaxy population, if anything even steeper than the evolution previously found for quasars and for infrared starburst galaxies. It seems odd in retrospect that earlier studies found no sign of this.

In another exciting development, Charles Steidel of the Palomar Observatory and collaborators have applied a technique which had previously been used in searches for high-redshift quasars. It was known that the emission from high-redshift galaxies would start to decline steeply at wavelengths shorter than that of Lyman alpha because such photons would be absorbed by gas either in the galaxy or in the intervening intergalactic medium. By taking images in three colour bands, ranging from ultraviolet to near infrared with 4-metre telescopes, they could pick out candidate galaxies with redshifts around 3 by their relative weakness in the ultraviolet image and then follow them up with spectroscopy on the Keck 10-metre telescope. They started to detect scores of galaxies with redshifts around 3 in 1995. Another technique which has started to yield dramatic results is to search for highly redshifted emission lines with narrow waveband filters targeted on the wavelengths of these redshifted lines.

## The Hubble Deep Field

The greatest impact on our view of the high redshift universe has come from a survey made with the Hubble Space Telescope (HST). In 1995 the Director of the HST, Bob Williams, made a very inspired decision to use his allocation of Director's Time, observing time which can be used for any programme of the Director's choice, to make a very deep survey of a small area of the sky. Some 150 orbits of the telescope were used to image a single small area about the size of Venus on the sky in four different colour bands in the ultraviolet, blue, green, and near infrared. The resulting combined colour image, the deepest view of the universe ever made, was released to the world in January 1996. The first time I saw it was in a BBC Television News studio, where I was called at very short notice to comment on this latest announcement from NASA. What caught my eye in the few minutes I had to study it before commenting were the numbers of brighter, peculiar-looking galaxies, presumably due to an increased incidence of interactions and mergers and the many very faint blue images, which I thought must be starburst galaxies. The Hubble Deep Field made an immediate and deep impression on the world's astronomers. Bob Williams' imaginative decision to release the image and

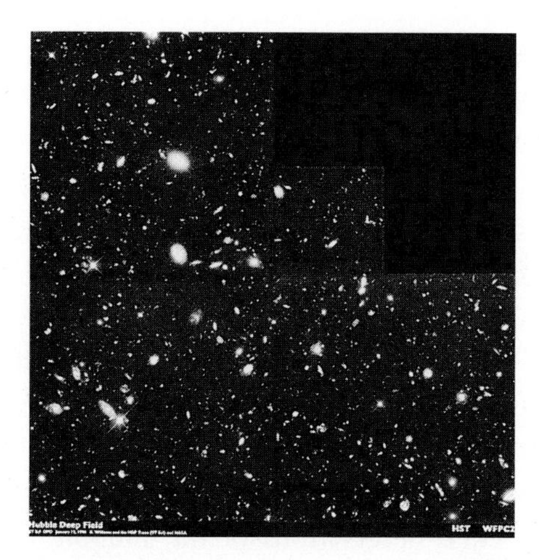


**Fig. 9.7** The Hubble Deep Field, the deepest image of the universe ever taken.

the data immediately was a stimulus for world-wide activity, both with other telescopes and by theoreticians.

I called together the extragalactic group at Imperial College and suggested that we focus our efforts for a few weeks on trying to understand the nature of the galaxies in the Hubble Deep Field. It was an exciting time and it soon became clear that many other groups around the world were engaged in a similar task. At Imperial we gradually focused on an attempt to estimate the redshifts of the galaxies in the survey, over 1600 of them in all, using the colour information given by the four images. Our main finding was that over half of the galaxies were at redshift greater than 2 and that many probably had redshift greater than 3, a conclusion reached independently by several groups. First into print were a group led by Amos Yahil of Stony Brook University, New York. This finding was confirmed in subsequent spectroscopic studies using the Keck telescope. A team at the Space Telescope Science Institute in Baltimore, led by Piero Madau, combined the results from the Hubble Deep Field with the earlier Canada–France Redshift Survey and attempted to put together a complete history of star formation in galaxies out to redshift 5. They concluded that the rate of star formation in galaxies had been at its peak at redshift 1–2, with a strong decline between redshift 1 and the present day, and indications of a decline between redshift 2 and 5. It seemed that we might be close to being able to see the whole history of star formation in galaxies from early times to the present.

## ISO observes the Hubble Deep Field

I was already leading a large European consortium of astronomers in a deep survey at mid and far infrared wavelengths with the Infrared Space Observatory (ISO), a 60-cm infrared telescope cooled with liquid helium to a temperature of –271 degrees Celcius (only 2 degrees above absolute zero; see p. 73). ISO was launched by the European Space Agency in November 1995 and observed the sky at wavelengths between 3 and 200 microns for nearly two and a half years. The goal of our survey, which is called the European Large Area ISO Survey (ELAIS) and from which the data are still being analysed, is to study the star formation history and look for high-redshift infrared galaxies. As soon as the HST observations of the Hubble Deep Field were released, my collaborators and I proposed to Martin Kessler, the Director of ISO, that the satellite be used to image the Hubble Deep Field at 6.7 and 15 microns, the most sensitive ISO wavelengths. Our idea was that if dust plays an important part in star-forming regions at high redshift, as it does in our Galaxy today, then we might see excess infrared radiation from some of the Hubble Deep Field galaxies. The small grains of dust tend to absorb ultraviolet and visible radiation and then re-radiate it at infrared wavelengths. Although several other groups made similar proposals we were lucky enough to be selected to design the observations and analyse the data. The condition was that we release the data to the community within three months of receiving them.

The ISO observations were made during July 1995 and we received the data at Imperial College over the next month. My team of postdoctoral researchers (Seb Oliver, Steve Serjeant, Pippa Goldschmidt, and Bob Mann) set to work to analyse the data, construct maps of the Hubble Deep Field at our two survey wavelengths of 6.7 and 15 microns, and look for sources of infrared radiation in the maps. To our delight we found a dozen infrared sources in our survey, which we could associate with galaxies with redshifts in the range 0.5–1, and for many the most natural interpretation was that the bulk of the energy from a starburst was being radiated at mid and far infrared wavelengths. The implied far infrared luminosities were high—once again we were dealing with ultraluminous infrared galaxies and very high rates of star formation. Our interpretation was nicely confirmed by a very deep radio survey of the Hubble Deep

Field carried out with the huge radio telescope known as the Very Large Array (VLA) in New Mexico by Ed Fomalont and colleagues from the US National Radio Astronomy Observatory (NRAO). For starburst galaxies there is a well-known correlation between the radio and far infrared outputs, so the rate of star formation can be deduced directly from the radio emission. The rates of star formation deduced by the NRAO team agreed well with those we estimated from the ISO data. It is clear that visible and ultraviolet observations can not reveal the total rate of star formation, since much of the energy from newly forming stars is absorbed by dust and radiated at mid and far infrared wavelengths.

A problem arose for us when a French group at Saclay, led by Catharine Cesarsky, reanalysed the ISO data from the Hubble Deep Field and concluded that they could not find all our sources, especially at 6.7 microns. To resolve this I asked Martin Kessler to re-observe the Hubble Deep Field with ISO at 6.7 microns. This was done and to my relief most of our sources were confirmed. One or two were spurious, as argued by the French group, but our conclusion that star formation rates were higher than expected remained valid.

There are not really enough infrared sources in the ISO survey of the Hubble Deep Field to make a definitive statement about the star formation rate and the fraction of starlight that is absorbed by dust. However, there have been a number of deep surveys made with ISO, including our own ELAIS survey, and when all these are analysed we

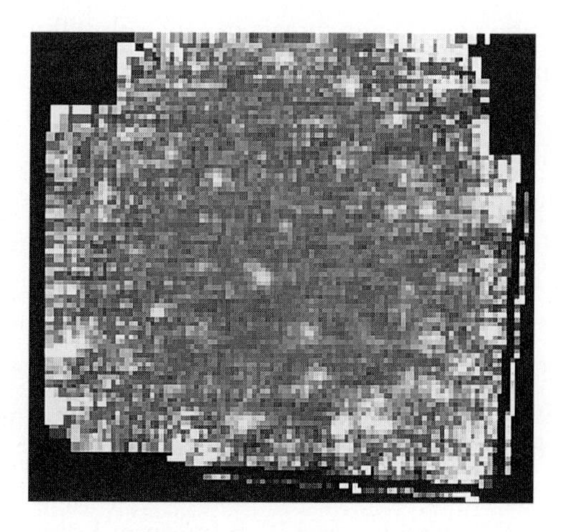

**Fig. 9.8** ISO image of the Hubble Deep Field at 15 microns.

will have a much more precise answer to these questions. Another very important test is being provided by deep surveys at submillimetre wavelength using the new 'SCUBA' instrument on the UK's James Clerk Maxwell Telescope on Hawaii. The Submillimetre Common User Bolometer Array (SCUBA) is the most powerful submillimetre instrument in the world and came into action during 1997. In anticipation of this moment I had organized a collaboration involving several of the leading UK groups working in submillimetre astronomy (Universities of Edinburgh and Cambridge, and Queen Mary and Westfield College, London, in addition to Imperial College) and we had applied for, and been awarded, a substantial block of time to make the first survey of the sky at submillimetre wavelengths.

It was very exciting to be at Mauna Kea, Hawaii, in January 1998, to embark on this survey. We had selected the Hubble Deep Field as the natural region for our first deep map, because it was so well studied at other wavelengths. It turns out that submillimetre wavelengths are especially sensitive to high-redshift starburst galaxies. The light from a galaxy at redshift 5 whose peak emission is at far infrared wavelengths, say 50–150 microns, would be redshifted to wavelengths of 300–900 microns, and this is just the range accessible from a high-altitude observatory like Mauna Kea, at 4000 metres. In July 1998 we published the first results from this survey, showing a map of the Hubble Deep Field at 850 microns with five reliably detected sources. Because the spatial resolution of SCUBA is not very good, so that we can only estimate the locations of the sources to a few seconds of arc, it is difficult to decide exactly which galaxies the submillimetre sources are connected with. However, we argued that most of these submillimetre galaxies have redshifts much larger than 1, probably in the range 2–5. The implication is that the galaxies are extremely luminous in the far infrared, in the ultraluminous class, and hence that the star formation rate at these redshifts is surprisingly high.

Soon after the start of our survey two rival surveys began, also using SCUBA. The first involved a collaboration between the University of Hawaii and Japanese astronomers, led by Len Cowie, and the second was a collaboration between Canadian and French astronomers, and astronomers from the University of Cardiff, led by Simon Lilly. In addition, a group of British astronomers (Ian Smail of Durham, Andrew Blain of Cambridge, and Rob Ivison of Edinburgh) have been mapping rich clusters of galaxies at 850 microns and using the gravitational lensing properties of the clusters to survey the volume behind the clusters for submillimetre galaxies. All these surveys detect 850-micron sources at about the same rate as ourselves and agree that many must be high-redshift galaxies forming stars at a prodigious rate, though there is some controversy about the fraction of sources at redshifts less than one.

If we put together all that we know about the star formation history from ISO and SCUBA surveys, and from ultraviolet surveys, but with careful correction for the dimming effect of interstellar dust, we find a consistent picture. At redshift 1, when the universe was a third of its present age, the star formation rate was 20 times higher than it is today. It increases by a further factor of two between redshift 1 and 3, when the universe was one eighth of its present age, and doesn't necessarily drop much between redshifts 3 and 5, at which time the universe was only 0.8 billion years old (7% of its present age). It seems that stars were already forming at a profuse rate when the universe was a billion years old. A picture of this kind is also consistent with everything we know about the numbers of faint galaxies at far infrared and

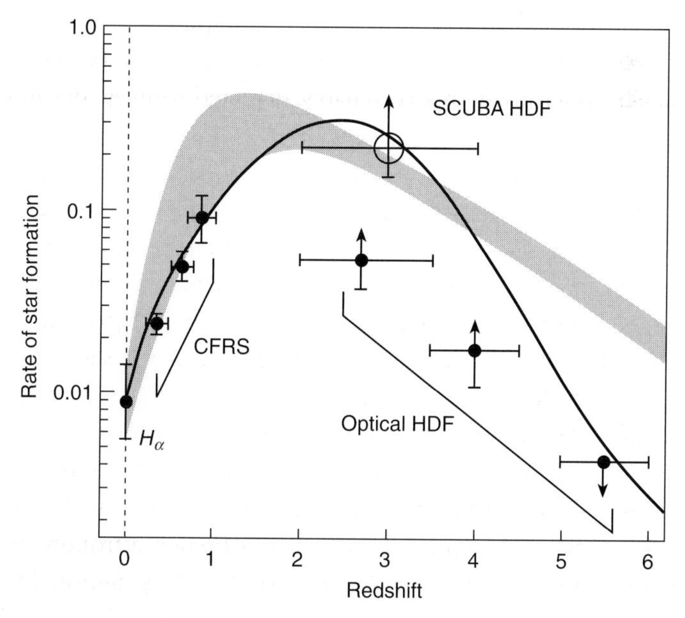


**Fig. 9.9** The star formation history of the universe, showing the high rate inferred at large redshifts from the UK SCUBA survey of the Hubble Deep Field.

submillimetre wavelengths and with measurements of the intensity of the submillimetre background radiation from galaxies made with the COBE satellite (p. 74).

## History of star formation: the ninth number

So the ninth number of the cosmos is the rate of evolution of starburst and active galaxies between now and the epoch corresponding to redshift about 2, which we interpret as the rate of decline of star formation over this period of time. If we model this as an exponential decline with time, then we can characterize this number by comparing the exponential time-scale, $\tau_{sf}$, to the age of the universe, $t_0$, so the ninth number is Q, where $\tau_{sf} = t_0/Q$. We find that Q is in the range 3–6, so that $\tau_{sf}$ is between 2 and 4 billion years. However, with all the new results from the Hubble Deep Field, the ground-based searches for high-redshift galaxies, and from the ISO and SCUBA surveys, we can now be more ambitious and try to characterize the whole history of star formation in galaxies. At the moment we probably can not do this with a single number. In addition to the exponential time-scale defined above, we probably need a second number to define the epoch at which the star formation in galaxies reached its peak, which seems to be between redshift 2 and 5, at least for spiral galaxies. Even these two numbers may not give us the whole history of star formation. For example, the history of how stars formed in elliptical galaxies remains obscure. It clearly happened long ago as almost no star formation goes on in these galaxies today. But did the stars in ellipticals, and in the central ellipsoidal bulges of spirals, form in single monolithic events at very early times, or can we think of them as just the end-point of particularly energetic starbursts from mergers which are part of the same story as that for spirals? We will need new deep surveys at optical and infrared wavelengths to resolve this question.

So it is simplistic to assign a single number to describe the history of star formation in galaxies. Much remains to be discovered observationally and many exciting developments can be expected in the near future from ISO and SCUBA. Further into the future there will be NASA's Space Infrared Telescope Facility (SIRTF), due for launch in 2001, the Japanese Astro-F far infrared survey mission, due for launch in 2003, and the European Space Agency's Far Infrared and Submillimetre Telescope (FIRST), due for launch in 2007, which will all focus on the search for high-redshift dusty galaxies in the process of formation. Theoretical models for how we get from the density fluctuations that we see in the microwave background to an actual galaxy of stars, gas, and dust remain in a rather primitive state. The physics of the star formation process, even in our own neighbourhood, remains unclear and needs many parameters to describe it. Eventually we might hope that this whole process of galaxy formation and evolution will be understood as a natural consequence of the initial density fluctuations, the nature of the dark matter in the universe, and the cosmological parameters. When we get to that stage we may not need to assign a separate cosmic number to this process at all.

Chapter 10

# The nine numbers of the cosmos

*Thus the explorations of space end on a note of uncertainty. And necessarily so. We are, by definition, in the very center of the observable region. We know our immediate neighbourhood rather intimately. With increasing distance, our knowledge fades, and fades rapidly. Eventually, we reach the dim boundary—the utmost limits of our telescopes. There, we measure shadows, and we search among ghostly errors of measurement for landmarks that are scarcely more substantial.*

*The search will continue. Not until the empirical sources are exhausted, need we pass on to the dreamy realms of speculation.*

Edwin Hubble, *The realm of the nebulae*, 1936

In this chapter we pull together what we know about the nine numbers of the cosmos, what the viable universe models are, how close we are to final determination of these numbers, and where the uncertainties and enigmas remain. We ask: how close are we to knowing the universe? And we look forward to the future: what can we expect astronomy and particle physics to tell us over the next 10 or 15 years? In that period we will have the launch of NASA's MAP (launch 2001) and ESA's PLANCK Surveyor (launch 2007) missions, both dedicated to mapping the microwave background radiation in ever finer detail. In the fine details of the microwave background fluctuations are recorded the imprints of all the main cosmological parameters. Especially after PLANCK we will know many of the cosmological parameters to great accuracy. We will also have ESA's Far Infrared and Submillimetre Telescope, FIRST, also being launched in 2007 on the same Ariane 5 launcher as PLANCK, which will open up this final unexplored 200–500 micron waveband and should help us to resolve some of the uncertainties about the history of star formation. We hope to have the Next Generation Space Telescope (NGST) in orbit by 2008, an 8-metre telescope which will work in the wavelength range 1–5 microns, and perhaps to longer wavelengths, and allow us to study starlight out to a redshift of 10. By 2010 we should have the Large Millimetre Array, an international project consisting of 64 12-metre telescopes providing spectacular angular resolution and sensitivity at millimetre and submillimetre wavelengths, which will provide a wonderful complement to FIRST and NGST in studying the high-redshift universe.

On the particle physics front we will have the switching on of CERN's Large Hadron Collider (LHC), due to start working in 2005, which will probably be capable of detecting the neutralino (p. 98), if it has not already been found in underground experiments, as well as many other exotic particles expected by the theorists. A variety of

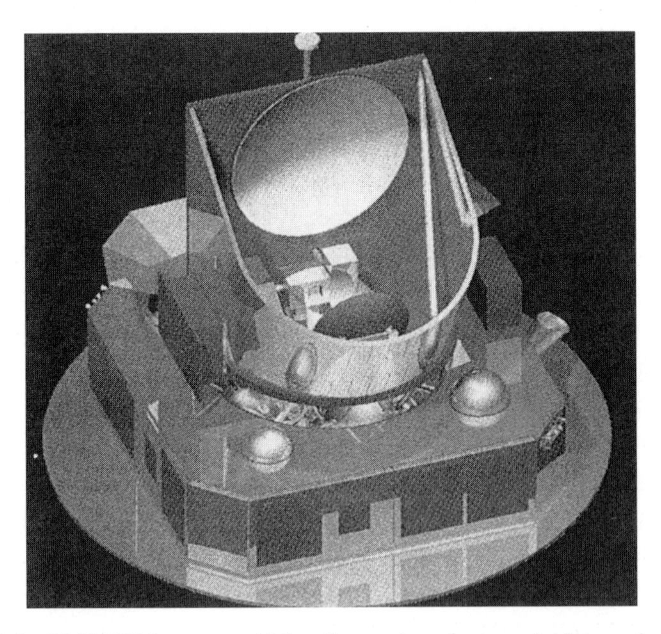

**Fig. 10.1** PLANCK Surveyor, which will map the microwave background radiation to unprecedented accuracy in 2007.

**Fig. 10.2** The Far Infrared and Submillimetre Telescope, FIRST, also due for launch in 2007.

neutrino experiments both underground and under the sea will probably determine the masses of all three neutrino types over the next decade. The underground dark matter experiments should have decisively determined the nature of the non-baryonic dark matter in the halo of our Galaxy.

Of course predicting the future of science is a hopeless task. There is a *Trivial Pursuit* question which goes: 'How long before the first Sputnik did the British Astronomer Royal say that space travel was bunk?' Answer: two years. This was Sir Richard Woolley in 1956, two years before the launch of Sputnik. In fact he kept on saying this even after the launch of the first piloted spacecraft because he was strongly sceptical of the feasibility of *interstellar* travel. So perhaps it is unjust that he is immortalized in *Trivial Pursuit*. Anyway, there are plenty of other precedents for scientists making confident statements about the future which are made to look ridiculous very quickly. With that caveat let me summarize my estimates for each of the nine numbers and try to predict how accurately they will be known in 2015, following the MAP, PLANCK, FIRST, and NGST missions, and the advent of the LHC:

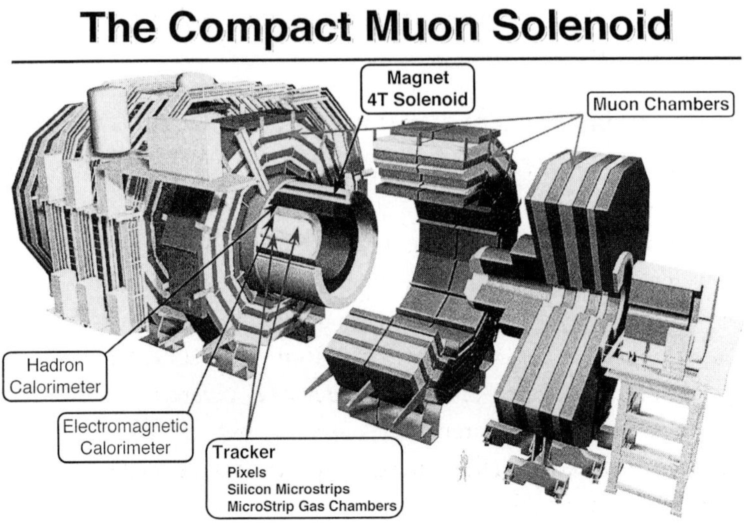

**Fig. 10.3** The Compact Muon Solenoid, which will be one of the main detectors for the Large Hadron Collider, or LHC, at CERN.

## 1 The density of baryonic matter, $\Omega_b$

The density of ordinary baryonic matter is quite well determined from the abundances of the light elements helium, deuterium, and lithium as (for $H_0 = 65$ km s$^{-1}$ Mpc$^{-1}$)

$$\Omega_b = 0.03 \pm 0.006, \text{ a 20\% accuracy.}$$

Searches for deuterium in the spectra of high-redshift quasars are already refining this number and by 2015 we can expect that PLANCK will have determined the baryonic density to an accuracy of 1%. We can also expect that we will have determined how much of this matter is in the form of visible stars, dead remnants, and intergalactic clouds of gas, at least out to redshift 5. As we are already beginning to peep into the universe at redshift greater than 5, with several galaxies now known with such redshifts, we can hope that enormous progress will be made by 2015 in understanding the redshift 5–10 era. Locally, the microlensing experiments should have measured the mass of dark baryonic objects in the halo of our Galaxy to high precision. So perhaps by 2015 we will know precisely not only how much baryonic matter there is, but also how it is distributed between stars, dead stellar remnants, brown dwarfs, and interstellar and intergalactic gas.

## 2 The anisotropy of the universe, $\Delta T/T$

The estimate from COBE for the anisotropy on large scales, which we characterized by $\Delta T/T$, is 1 part in 100 000, that is $1.0 \times 10^{-5}$, to an accuracy of 25%. Following MAP and PLANCK the accuracy with which this is known will improve slightly to about 10%. On the other hand the accuracy of the anisotropy on small scales (at present not really known at all) will, particularly from PLANCK, be known to very high accuracy, to about 1%. Because different cosmological scenarios make very different predictions about the anisotropies on small scales, this in turn allows many other cosmological parameters to be determined precisely.

## 3 The Hubble constant, $H_0$

The current best estimate for the Hubble constant, which measures the rate of expansion of the universe, is $H_0 = 65$ km s$^{-1}$ Mpc$^{-1}$, with an uncertainty of about 12%. This is still capable of packing some surprises but a value outside the range 50–80 seems highly unlikely. In 2015 the value should be known, from the PLANCK mission, to within a few %, but methods like the gravitational lens time delay method (p. 53) could have delivered an accurate answer well before then.

## 4 The age of the universe, $t_0$

For the age of the universe, I have adopted the value $t_0 = 12$ billion years, with a total uncertainty of 2 billion years either way. This is a 20% reduction from the value of 15 billion years which was believed only a few years ago, a reduction brought about by the improved accuracy of the estimates of distances to globular clusters, where the oldest stars reside, made by the HIPPARCOS satellite team. The age of the universe is again a quantity capable of surprising developments and values in the range 10–15 billion years are still possible. This may be the hardest of the nine numbers for us to dramatically improve the accuracy with which it is known. No new precise methods of direct measurement are on the horizon and for once PLANCK will not help us. Since no new astrometric mission is currently approved to follow

HIPPARCOS, there may not be significant improvements in local distance measurements by 2015. However, there are already plans for successor missions (FAME, GAIA) and one of these may fly before 2015 and lead to improvements in the estimates of distances to globular clusters and hence of the age of our Galaxy. We can also hope that there will be improvements in understanding the stellar evolution of both globular cluster stars and white dwarfs, and improvements in analysis of radioactive decay estimates. Our improving understanding of the star formation history of galaxies should remove one of the uncertainties in the latter estimates. I estimate that the age of the universe will be known to 5% by 2015, an improvement of a factor of three on our present situation. Remember that we are talking about the age of the universe back to the Planck time and it is impossible to say what the history of the universe was prior to that. The lifetime of the universe in the future is either infinite or probably so large as to be virtually unmeasurable.

## 5 The temperature of the microwave background, $T_0$

COBE has already determined the temperature of the microwave background, $T_0$, as 2.728 degrees Kelvin to an accuracy of 0.1%. This is the one cosmic number which we really do know accurately. There are no plans to try to measure this to greater accuracy between now and 2015. Perhaps in the middle of the next century the interesting scientific programmes will be experiments to achieve an additional decimal place in the accuracy of the fundamental numbers of cosmology like $T_0$.

## 6 The density of cold dark matter, $\Omega_{cdm}$

There is enormous uncertainty in the density of cold dark matter, $\Omega_{cdm}$, with possible values in the range 0–0.97. Currently favoured values are in the range 0.2–0.4 and values lower than 0.1 seem unlikely. However, I certainly do not think anyone should bet their house against a universe with a critical density, $\Omega_{tot} = 1$, mainly composed of cold dark matter. If $\Omega_{tot} = 1$, then $H_0 = 65$ km s$^{-1}$ Mpc$^{-1}$ requires an age for the universe $t_0 = 10$ billion years, or alternatively if $t_0 = 12$ billion years, we require $H_0 = 55$ km s$^{-1}$ Mpc$^{-1}$. Neither of these alternatives seems beyond the realm of possibility at present. In 2015 we should know $\Omega_{cdm}$ and $\Omega_{tot}$ to an accuracy of 1%.

## 7 The tilt, the string tension, or the density of hot dark matter, $\Omega_{hdm}$

The missing ingredient to make galaxy formation and clustering work could be hot dark matter, a tilted primordial density fluctuation spectrum, cosmic strings, or some as yet unknown ingredient. If hot dark matter plays a key role, the likely value of the density of hot dark matter, $\Omega_{hdm}$, is about 0.2. The recent discovery of muon neutrino oscillations tells us that $\Omega_{hdm}$ is greater than or equal to about 0.001. So we seem to be in the frustrating position of demonstrating that hot dark matter exists but finding that it is present in such small amounts that it has no cosmological effect. Current and planned particle physics experiments are capable of determining the masses of the three neutrino types with sufficient accuracy to determine $\Omega_{hdm}$ over the next ten years.

Since inflationary models can accommodate an immense variety of primordial density fluctuation spectra it is difficult to be precise about the 'tilt' at this time. A change in the power spectrum index by 10%, well within the measured uncertainties, seems to be enough to do the trick for a pure cold dark matter scenario. It looks as if we have to wait for the PLANCK mission to tell us what the primordial density fluctuation spectrum was and then see what we can deduce about inflation and the early universe. Strings and other defects remain a serious possibility, though theorists have found it difficult to fit the observed distribution of galaxy clustering within these scenarios. Particle physicists believe that there have been several major phase transitions during the universe. Any of these could have left defects behind. The issue is: were they prevalent enough to affect the dynamics of the universe? MAP should make progress on this question and PLANCK should certainly settle it.

## 8 The cosmological constant, $\Lambda$

The cosmological constant is another largely unknown number, with the dimensionless value $\Lambda$ lying anywhere in the range 0 to 0.7. The high-redshift supernova surveys are claiming that $\Lambda$ has to be positive and that values towards the upper end of the permitted range are preferred. This may be a bit premature and we might still need to understand the physics of supernova events at earlier epochs better before we can be sure about these results. The statistics of gravitational lenses, which give us the limit of 0.9, could whittle away at the upper end of the range over the next few years. We may have to wait for PLANCK to give us a really precise value or limit. In 2015 we will either know $\Lambda$ to 1% (if it is of order 0.5–0.7) or have set a limit of $< 0.1$. $\Lambda = 0.9$, $\Omega_0 = 0.1$, $H_0 = 65$ km s$^{-1}$ Mpc$^{-1}$ would give an age for the universe of 19.3 billion years, which is almost certainly too high. On the other hand $\Lambda = 0.7$, $\Omega_0 = 0.3$ would give an age of 14.5 billion years for the same Hubble constant, so this is a possible, or even attractive, combination of parameters. This age would be slightly on the high side of the most recent estimates, but as we saw above, the age estimates are not very precise and may still change substantially.

## 9 The star formation history of the universe

We saw that the ninth 'number' is at present probably at least two numbers in practice. However, it is unlikely that both the seventh and eighth numbers are non-zero, so the total set of cosmic numbers is still nine. We are reasonably certain that the rate of star formation increases strongly as we look back towards the past, between the present epoch and redshift 2, with the rate of evolution characterized by the parameter $Q$, which is the ratio of the age of the universe to the exponential time-scale, having a value in the range 3–6. Beyond redshift 2 there is controversy. Evidence from the optical and ultraviolet is claimed to demonstrate that the rate of star formation decreases beyond redshift 2, while infrared and submillimetre surveys point to the peak star formation rate being sustained back to redshift 5 at least. The FIRST mission should resolve some of these questions. In 2015 I would expect that we know this function to an accuracy of 20%.

## Summary

The first four of our cosmic numbers are currently known to an accuracy of 15–25%. The fifth number is already known to an accuracy of 0.1%. The last four are hugely uncertain, since they deal with the mysteries of dark matter, the inflationary era (if it existed), and the origin of density fluctuations and their evolution into galaxies and stars. Looking ahead to 2015 I am predicting that most of these cosmic parameters will be known to an accuracy of 1%. First hints of the values of some of the parameters will come from the MAP mission but the greatest advance in precision will come from Europe's PLANCK

Surveyor mission. The whole science of observational cosmology will have shifted into much more detailed study of the process of formation and evolution of galaxies. Astronomers will be using new generations of space-based and ground-based telescopes to study the universe at high redshift and it is reasonable to suppose that the emphasis will have shifted from the redshift range 1–5 currently under scrutiny to much earlier epochs when galaxies were just beginning to assemble and form the first stars. The FIRST telescope, due to be launched into orbit in 2007, will search for high-redshift, star-forming galaxies. The Next Generation Space Telescope (NGST), the successor to the Hubble Space Telescope, and with at least twice its diameter, should be in action by 2008. And on the ground we can expect to have the Large Millimetre Array, capable of detecting very faint millimetre sources and locating them with immense precision, by 2010. In the next few years there will be over a dozen 8-metre-class optical telescopes in action around the world. With this repertoire of telescopes and missions we can hope by 2015 to have understood and quantified most of the directly observable universe back to a redshift of 10, and to have significant insight into the dark epoch back to the era of recombination at redshift 1000.

Particle physicists, on the other hand, will be able to take the spectrum of primordial density perturbations measured by PLANCK and deduce whether and in what form inflation took place. With the advent of the Large Hadron Collider they will hope to have detected Higgs particles, the key to understanding why elementary particles have mass, and the lightest supersymmetric particle, the neutralino. Confirmation of the idea of supersymmetry would be an important step towards unification of forces, both Grand Unification of the strong nuclear force with the electroweak force, and the more demanding unification of quantum theory and gravity. Superstring theorists believe they are on course for arriving at a successful merging of these two fundamental but apparently irreconcilable theories. Perhaps we can hope for some insight into whether the cosmological constant should be non-zero today. We can hope that astronomical studies of large-scale structure and of the microwave background fluctuations should have established whether topological defects left over by phase transitions play a major role in the universe. By 2015 we should at least expect to know whether neutrinos have mass and whether the simplest cold dark matter candidate, the neutralino, exists, so that much of our ignorance about our cosmic numbers six, seven, and eight might be resolved. We should have an accurate picture of the evolution of the universe back to a time one million millionth of a second after the Big Bang, that is $10^{-12}$ seconds.

Even if we knew all the nine cosmic numbers to perfect accuracy, we would still not know what happened in the very first instants of the universe. I wonder whether anything that happened before the Planck time, $10^{-43}$ seconds after the Big Bang, is knowable. Theoretical particle physicists love to speculate about this era because it stretches their theories and their techniques to the limit. But the prospect of testing these ideas against reality does not seem good. It will be a great achievement if theorists can come up with a self-consistent quantum theory of gravity (see p. 84). But if the only predictions of such a 'Theory of Everything' relate to the Planck era it will remain a metaphysical and untestable theory. Superstring theory and its rivals are a necessary intellectual exercise, but may not lead to any practical advance in physics or cosmology.

Our knowledge is limited by the size of accelerators and nuclear reactors that can be created on earth and by the size of telescopes that we can build on the ground or in space. Astronomers and particle physicists have shown immense ingenuity and imagination in unravelling the cosmological story within these constraints. There are still new types of telescope yet to make their full impact on cosmology:

## Neutrino telescopes

These are already operating—the most famous being the Japanese Kamiokande detector (p. 81)—and have detected neutrinos from the sun and from supernova 1987A in the Large Magellanic Cloud. Future generations of neutrino telescope might be capable of detecting the cosmological background of neutrinos, penetrating the fog of the radiation-dominated era. Neutrinos could allow us to probe back to 1 second after the Big Bang before they, too, join the fog of particles in equilibrium with radiation.

## Cosmic ray observatories

The highest energy cosmic rays have energies thousands of times the energy of the proton and could allow us to probe physics at energies beyond those that can be achieved by terrestrial accelerators, thereby allowing us to understand conditions at even earlier epochs in the universe. Future planned cosmic ray observatories like AUGER have the potential to take us to these energies.

### Neutralino telescopes

The underground cold dark matter experiments now in operation (p. 98) may be the first prototypes of neutralino telescopes which could eventually probe very early times in the universe. We can imagine that in the next century there will be neutralino telescopes deep underground studying the cosmic neutralino background generated $10^{-4}$ seconds after the Big Bang.

### Gravitational wave observatories

Gravitational waves are one of the major predictions of general relativity yet to be directly tested experimentally. They manifest themselves by a faint stretching and compression of space and time as they pass. It is hoped that they may be detected by monitoring the distance between two small masses, usually mirrors between which laser beams are bounced. Indirect evidence for gravitational waves has been found by studying the slowing down of the orbital speed of the binary pulsar.

A series of gravitational wave observatories are now being constructed or planned: the Japanese TAMA 300 in Tokyo (1999), the German–British Geo 600 in Hanover (2000), the US LIGO (2000–2001), the French–Italian VIRGO (2002). These experiments hope to make the first direct detection of gravitational waves, from rotating neutron stars undergoing an instability, from neutron star mergers, or even from the merger of massive black holes in the nuclei of galaxies. Only in very extreme scenarios, like cosmic string models, can these experiments hope to detect a cosmological background of gravitational waves from the early universe. The same applies to the more ambitious Laser Interferometer Space Antenna (LISA) mission, a four-satellite gravitational wave observatory proposed for launch by the European Space Agency in 2010–15. LISA should certainly detect gravitational waves from many types of astronomical object, including binary systems containing neutron stars, but again could only detect cosmological gravitational waves in very extreme scenarios. An improvement in sensitivity by a further factor of 1000 will be needed to detect gravitational waves generated in the more standard inflationary models and this may take a further 20–30 years to achieve. But while PLANCK's map of the microwave background fluctuations should show us in detail the spectrum of density fluctuations that emerged from the inflationary era (if this existed), the gravitational wave observatories of 2050 will allow us to probe the inflationary era itself.

## Epilogue

The twentieth century has been a time of extraordinary scientific advances and this is nowhere so clear as in cosmology, where almost everything we know about the universe beyond our Galaxy has been learnt since 1900. The next 10 or 20 years should see the solution of the problem of the observable universe and the determination of all the nine numbers of the cosmos. Beyond that we have to find ways to make the unobservable early universe visible to us. Human imagination being what it is there may be no limit to what we can achieve. It would be foolish to say that cosmology will come to an end in the next century. On the contrary, there will be whole new areas of cosmological science that we can not yet imagine. I make only one prediction for the year 2100, based on the practical limitations that exist to the size of telescopes and accelerators that can be built, and to the resources that can be spent on scientific experiments: that the Planck era, and what, if anything, preceded it, remains shrouded in mystery. It would not surprise me if this were still a mystery in the year 3000.

# Further reading

Hubble, Edwin, *The Realm of the Nebulae*, 1936, Yale University Press/Dover

Lemaître, Georges, *The Primeval Atom*, 1950, Van Nostrand

Gamow, George, *The Creation of the Universe*, 1957, Viking

Weinberg, Steven, *The First Three Minutes*, 1977, Basic Books

Rowan-Robinson, Michael, *The Cosmological Distance Ladder*, 1985, W. H. Freeman

Dawkins, Richard, *The Blind Watchmaker*, 1986, Longman

Gould, Stephen Jay, *Life's Grandeur/Full House*, 1986, Harmony Books/Random House

Harrison, Edward, *Darkness at Night*, 1987, Harvard University Press

Gould, Stephen Jay, *Wonderful Life*, 1989, Hutchinson/Penguin

Schramm, David, and Michael Riordan, *The Shadows of Creation*, 1990, W. H. Freeman

Smoot, George, and Keay Davidson, *Wrinkles in Time*, 1993, William Morrow

Rowan-Robinson, Michael, *Ripples in the Cosmos*, 1993, W. H. Freeman/Spektrum

Rowan-Robinson, Michael, *Cosmology*, 3rd edition 1996, Oxford University Press

Mather, John, and John Boslough, *The Very First Light*, 1996, Basic Books

# Glossary

**absolute zero** the temperature at which random motions of atoms cease (–273 degrees Celsius), taken as the zero point for the Kelvin (K) scale of temperature.

**anisotropy** of universe: deviation of the distribution of galaxies or radiation from isotropy (q.v.).

**annihilation** when a particle and its antiparticle (e.g. electron and positron) collide and destroy each other, leaving only pure energy in the form of photons (q.v.).

**antimatter** matter composed of particles which have the same masses as normal matter particles but opposite charge and spin.

**antiparticle** for every particle there is an antiparticle that has the same mass but the opposite charge and spin (except the photon, which has no charge, mass, or spin, only energy).

**atomic number** the number of protons (and electrons) in an atom.

**atomic weight** the mass of the atom of an element relative to hydrogen.

**axion** a candidate cold dark matter particle required in certain particle physics models.

**baryon** heavy fundamental particle such as the proton and neutron (q.v.).

**baryonic matter** normal matter composed of neutrons and protons, in contrast to hot and cold dark matter (q.v.).

**Big Bang** the initial moment of expanding universe models, which mathematically appears as the instant at which density and other physical properties become infinite. In practice cosmological models can not be extrapolated back before Planck time (q.v.).

**Big Crunch** the final moment of 'oscillating' universe models in which the density of the universe is high enough to halt expansion and turn it into a collapse. As with the Big Bang (q.v.) we can not follow this collapse through to the predicted infinite density and can get no closer to it than the Planck time.

**billion** one thousand million.

**black body** a perfectly efficient absorber or emitter of radiation. It has the characteristic Planck spectrum peaking at wavelengths depending only on temperature. A black body spectrum is the signature of a gas in which the matter and radiation are in thermal equilibrium.

**black hole** a region from which the escape speed exceeds the speed of light, so no matter or signal can escape. The black hole mass is enclosed within an event horizon and is invisible, though we still feel its gravitational effects. Formed as the end-point of the evolution of a very massive star (> 20 times the mass of sun) or in the nuclei of galaxies.

**brown dwarf** a gaseous, self-gravitating object in the mass range 1–80 times that of Jupiter, which is too massive to be considered a planet and of too low a mass for nuclear reactions to ignite and make a star.

**Cepheid variable star** a massive pulsating star whose light output varies regularly on a time-scale of 1–100 days. The key property is that the luminosity is related to the period, so Cepheids can be used as distance indicators.

**cold dark matter** the postulated form of non-baryonic dark matter in which particles (e.g. neutralino or axion (q.v.)) are moving slowly in the early universe.

**Copernican principle** the principle that the earth is not in a special place in the universe.

**cosmic string** a type of topological defect (q.v.) consisting of string-like regions of very high energy density left over from the early universe.

**cosmological constant** the parameter characterizing the cosmological repulsion (q.v.).

**cosmological principle** Einstein's hypothesis that the universe is homogeneous and isotropic (q.v.).

**cosmological repulsion** the additional force, introduced by Einstein, whose effect increases with distance. The particle physics interpretation is that it represents the energy density of the vacuum.

**critical density** the density of the universe that separates models which recollapse from those that keep on expanding monotonically.

**dark matter** matter whose existence is inferred only from its gravitational effects

**defects** see *topological defects.*

**density fluctuation/perturbation** a region in which the density of the universe is slightly above or below the average.

**density fluctuation spectrum** how the number of fluctuations of different masses varies with mass.

**density parameter** the dimensionless measure of the density of the universe, the ratio of the density to the critical density.

**dipole anisotropy** the effect of our Galaxy's peculiar motion through the cosmic frame, resulting in the microwave background temperature appearing slightly higher in the direction of motion, and slightly cooler in the opposite direction.

**Doppler shift** the shift of wavelength or frequency caused by relative motion of source of radiation and observer. A source of optical radiation moving away from us is shifted towards the red end of the spectrum. Most galaxies are seen to be redshifted because of the expansion of the universe.

**Einstein–de Sitter model** model of the universe with critical density (q.v.).

**electron** the basic constituent of the atom with negative charge and 1/1836 times the mass of the proton. The cloud of electrons orbiting an atom determine its chemical properties.

**electroweak force** unification of the electromagnetic force and the weak nuclear force (q.v.). The electroweak symmetry is broken and the electroweak force separates into its constituent forces about $10^{-11}$ seconds after the Big Bang.

**entropy** the measure of disorder in a thermodynamic system.

**Grand Unified Force** postulated unification of electroweak and strong nuclear forces (q.v.). The Grand Unified Symmetry is believed to be broken and the Grand Unified Force to separate into its constituent forces about $10^{-35}$ seconds after the Big Bang.

**gravitational lens** light from a background source is bent round a star, galaxy, or cluster of galaxies (an effect of general relativity), causing magnification of the source and its break-up into multiple images, arcs, or a ring.

**hadron** heavy nuclear particles (baryons (q.v.) and mesons) which take part in strong nuclear interactions.

**hadron era** early era of the universe, prior to one millionth of a second after the Big Bang, when quarks, leptons, and their antiparticles are all in equilibrium with radiation.

**Harrison–Zeldovich spectrum** simple form of primordial density fluctuation spectrum (q.v.) in which the amplitude of density fluctuations scales as the inverse square of their size.

**heavy elements** the elements from carbon onwards (i.e. excluding hydrogen and the light elements (q.v.)).

**homogeneity** of the universe: the universe looks the same at every location.

**horizon** limit to the size of the observable universe, set by the finite age of the universe and the fixed speed of light.

**hot dark matter** non-baryonic matter particles which move around close to the speed of light in the early universe, for example neutrinos with a non-zero mass.

**Hubble constant** the constant of proportionality in Hubble's velocity–distance law, $H_0$, measured in km $s^{-1}$ $Mpc^{-1}$ (so has dimensions of 1/time).

**Hubble law** velocity is proportional to distance, implying an expanding universe.

**Hubble time** the inverse of the Hubble constant, $\tau_0$.

**inflation** the phase in the early universe, probably as result of phase transition (q.v.), during which the universe undergoes exponential expansion, driven by the energy density of the vacuum.

**isotope** of a given element has the same number of protons in its atomic nucleus as the element, but additional numbers of neutrons.

**isotropy** of the universe: the universe looks the same in every direction.

**large-scale structure** large-scale distribution of galaxies is characterized by clusters, voids, sheets, and filaments, but tends to look smoother as we look to ever larger scales.

**lepton** light sub-atomic particles, for example electrons and neutrinos.

**light elements** helium, lithium, beryllium, and boron.

**microlensing** gravitational lensing (q.v.) by stars magnifies brightness of background stars, but multiple images are on too small a scale to resolve.

**microwave background radiation** background radiation, discovered at microwave wavelengths in 1965, which is a relic of the early radiation-dominated phase of the universe.

**mixed dark matter** dark matter model involving both hot and cold dark matter (q.v.).

**monopoles** topological defects (q.v.), which are points of extremely high energy density possibly left over after early universe phase transitions.

**muon** a short-lived, light atomic particle which takes part in weak nuclear interactions.

**neutralino** lightest of additional particles postulated by supersymmetry (q.v.), and favoured candidate for cold dark matter.

**neutrino** particles found to be emitted in radioactive $\beta$-decay of neutron. There are three types of neutrino: the electron neutrino, the muon neutrino, and the tau neutrino. In the standard model of particle physics neutrinos have zero mass, but many physicists believe that neutrinos have a small non-zero mass.

**neutron** a heavy, uncharged sub-atomic particle, which, with the proton, is the fundamental constituent of atomic nuclei.

**neutron star** dead star, left as a remnant after a supernova explosion, in which the pressure of neutrons holds the star up against gravity.

**nucleon** nuclear particle, for example neutrons and protons.

**peculiar velocity** random motion of a galaxy about the Hubble law, generated by attraction of other galaxies and clusters of galaxies.

**phase transition** change in fundamental properties of matter, for example conversion of water to ice. In the early universe, phase transitions are expected to be associated with the symmetry breaking of the Grand Unified Force into the strong nuclear force and the electroweak force, and of the electroweak force into the weak nuclear force and the electromagnetic force.

**photon** particle of light, which has no spin, charge, or mass, but only energy.

**Planck time** period very shortly after the Big Bang (or shortly before the Big Crunch) when the validity of the general theory of relativity breaks down and a quantum theory of gravity is required.

**positron** the antiparticle of the electron with the same mass but opposite charge.

**proton** heavy, positively charged sub-atomic particle, which, with the neutron, is the fundamental constituent of atomic nuclei.

**quantum theory of gravity** to describe the universe very close to the Big Bang (or Big Crunch), we would need a new theory which integrates quantum theory and general relativity. The best candidate at the moment seems to be superstring theory (q.v.).

**quark** the building block of neutrons, protons, and other hadrons.

**radiation dominated phase** the early phase of the universe when radiation is the dominant form of energy.

**radioactive decay** comes in three types: $\alpha$-decay is the splitting (fission) of an atomic nucleus with the ejection of a helium nucleus; $\beta$-decay is the decay of a neutron into a proton with the ejection of an electron and an (electron) neutrino; $\gamma$-decay is the emission of a high-energy photon of light.

**recombination era** the moment when the temperature of the universe drops low enough (3000 K) for electrons to combine with protons to make neutral hydrogen atoms, leaving the universe transparent to radiation.

**redshift** see *Doppler shift*.

**relativity, general theory of** Einstein's theory of gravity, in which space–time is curved, and light is bent around masses.

**singularity** a region at the centre of a black hole or at the initial instant of the Big Bang, where general relativity predicts an infinite density of matter.

**spin** quantized property of elementary particles, analagous to the classical concept of spin about an axis.

**standard error** when a series of measurements are made of a physical quantity, the spread in the results is characterised by the standard error (the root-mean-square deviation of the measurements from the mean value).

**standard model of particle physics** consists of the Salam–Weinberg theory of the electroweak force (q.v.) plus the 'quantum chromodynamics' theory of quark interactions.

**steady state model** model of the universe in which, in addition to being homogeneous and isotropic, the universe is unchanging with time.

**strong nuclear force** the short-range force which governs nuclear reactions and nuclear structure.

**Sunyaev–Zeldovich effect** the free electrons in the very hot gas in clusters of galaxies boost the energy of microwave background photons to higher energy, changing the background spectrum in the direction of the cluster.

**superstring theory** the attempt to unify all forces of physics into a single mathematical structure, involving at least ten dimensions.

**supersymmetry** the theory which postulates additional symmetries to bring quarks and leptons onto the same basis and requires additional particles to exist, of which the lightest is the neutralino.

**tau** a short-lived light atomic particle which takes part in weak nuclear interactions.

**tilt** a modification of the primordial density fluctuation spectrum from the Harrison–Zeldovich form (q.v.) to account for excess structure seen on large scales.

**topological defects** regions of very high energy density in the form of points (monopoles), lines (strings), or plane surfaces (domain walls), left over after phase transitions in the early universe.

**weak nuclear force** the force which governs $\beta$-decay, and other interactions involving leptons (q.v.).

**white dwarf star** a dead star left as a remnant after the death of a star like the sun, in which the pressure of electrons holds the star up against gravity.

**WIMPs** weakly interacting massive particles believed to make up the bulk of the dark halo of our Galaxy, for example the neutralino (q.v.).

# Name Index

# Subject Index